Table of Contents

1 Introduction to Bayesian Methods

Bayes' theorem, at its core, is a mathematical formula that elegantly describes how to update our beliefs in light of new evidence. It's not just a theoretical concept; it's a powerful tool for making informed decisions, especially in situations rife with uncertainty, such as those commonly encountered in bioinformatics. The theorem itself is relatively straightforward: $P(A|B) = [P(B|A)P(A)] / P(B)$, where $P(A|B)$ represents the probability of event A happening given that event B has already occurred. Understanding this equation is crucial, as it forms the bedrock of all Bayesian inference. Its practical application, however, extends far beyond a simple formula, influencing everything from predicting gene expression to identifying disease markers. Mastering this fundamental concept unlocks the door to a wealth of sophisticated bioinformatics techniques.

The beauty of the Bayesian approach lies in its ability to incorporate prior knowledge. Before we even look at the data, we might have some preconceived notions, some prior beliefs about the phenomenon we are studying. These beliefs are formalized as a prior distribution, representing our uncertainty about the parameters of interest before observing any data. This contrasts sharply with frequentist methods, which often ignore prior information.

For instance, if we're studying a gene known to be involved in a particular metabolic pathway, we might assign a higher prior probability to it being differentially expressed in a related experiment. This prior knowledge, intelligently integrated, helps to refine our estimates and can dramatically improve the accuracy of our analyses.

The process of Bayesian inference is iterative. As we gather more data, we refine our initial beliefs. This refinement is achieved by combining the prior distribution with the likelihood function, which describes the probability of observing the data given specific parameter values. The result of this combination is the posterior distribution, representing our updated beliefs after considering the evidence. This cyclical nature—prior, likelihood, posterior—is the essence of Bayesian updating. It's a continuous learning process, where our understanding of the system evolves as we obtain more information. This contrasts with the one-shot nature of many frequentist techniques.

One significant advantage of Bayesian methods is their ability to quantify uncertainty. Unlike frequentist approaches which often provide point estimates, Bayesian inference provides a full posterior distribution. This distribution encapsulates all possible values for the parameters of interest, along with their associated probabilities. This offers a much richer understanding of the system, allowing us to not just pinpoint the most likely value, but also assess the range of plausible values and the associated uncertainty. In bioinformatics, where data is often noisy and incomplete, this ability to handle uncertainty explicitly is invaluable for generating robust and reliable conclusions. For example, when analyzing microarray data, we can not only estimate the fold change in gene expression, but also quantify our confidence in that estimate.

Another key aspect is the choice of prior distribution. This choice reflects our subjective belief and influences the results. While this might seem like a drawback, it's actually a strength. The ability to explicitly incorporate prior knowledge allows us to personalize our analyses. A carefully chosen prior can improve the efficiency of the inference process, especially with limited data. However, a poorly chosen prior could bias the results. The field offers various strategies for selecting priors, from informative priors reflecting existing knowledge to uninformative priors that minimally influence the posterior. Choosing the appropriate prior is a crucial step in any Bayesian analysis, influencing the accuracy and efficiency of inferences. Careful consideration and justification of prior selection are therefore essential components of a rigorous Bayesian analysis. Ultimately, the choice depends on the available information, the context of the problem, and the researcher's judgment.

Practical implementation of Bayesian methods often involves Markov Chain Monte Carlo (MCMC) techniques. These sophisticated algorithms allow us to sample from complex posterior distributions, even when analytical solutions are unavailable. MCMC methods generate a sequence of samples that, in the long run, approximate the true posterior distribution. This allows us to estimate various quantities of interest, such as posterior means, credible intervals, and probabilities of different hypotheses. While understanding the theoretical underpinnings of MCMC is beneficial, many user-friendly software packages abstract away the technical complexities, allowing even beginners to apply these powerful tools effectively. Software choices range from general-purpose statistical packages to specialized bioinformatics tools. Each package offers a slightly different interface and functionality. The choice depends on one's familiarity with the interface and the specific requirements of the analysis.

In summary, Bayesian methods offer a powerful and flexible framework for analyzing biological data. By incorporating prior knowledge and quantifying uncertainty, they allow for more robust and informative inferences. While the underlying mathematics might seem daunting at first, the practical application is often streamlined by readily available software packages. This chapter provided a foundational understanding of Bayes' theorem and its application in estimating parameters and performing statistical inferences. The following chapters will build upon this foundation, exploring specific applications of Bayesian methods in various bioinformatics domains. Understanding this foundation is crucial for effectively using and interpreting the results of Bayesian methods in the subsequent analyses presented in this book.

1.1 Bayes' Theorem and its Applications

Bayes' Theorem, at its core, provides a mathematical framework for revising our beliefs in light of new evidence. It elegantly combines prior knowledge, represented by a prior probability distribution, with observed data, expressed through the likelihood function, to yield a posterior probability distribution. This posterior distribution reflects our updated understanding, incorporating both prior beliefs and the information gleaned from the data. The theorem itself, expressed as P(A|B) = [P(B|A)P(A)] / P(B), might seem intimidating at first glance, but its application is surprisingly intuitive and powerful, especially in bioinformatics.

Consider the task of identifying a specific gene mutation associated with a particular disease. We might start with some prior belief about the prevalence of this mutation in the general population, perhaps based on previous studies. This forms our prior probability. Then, we analyze genomic data from patients with the disease. The likelihood function quantifies how probable the observed data (the presence or absence of the mutation) is, given that a patient has the disease. By combining these two elements—our prior belief and the likelihood from the data—Bayes' Theorem provides us with a refined, posterior probability of the mutation's presence in a disease-affected individual. This revised probability is significantly more informative than relying solely on the observed data or our prior assumptions alone.

Furthermore, the iterative nature of Bayesian inference is particularly valuable. The posterior probability from one analysis can serve as the prior for the next, allowing for continuous refinement of our understanding as more data becomes available. Imagine incorporating data from multiple patients, or even different studies. Each new dataset updates the posterior probability, leading to a progressively sharper and more accurate estimation of the mutation's prevalence and association with the disease. This iterative process is not only efficient but also reflects how scientific understanding evolves – continually building upon previous knowledge and integrating new findings.

The beauty of this approach lies in its ability to incorporate prior biological knowledge, something often overlooked by frequentist methods. For instance, we might have prior information about the location of certain genes on a chromosome or the functional relationships between proteins. Bayes' Theorem allows us to formally incorporate this information into our analysis, leading to more robust and biologically relevant conclusions. This is especially crucial in fields like bioinformatics where prior knowledge can significantly improve the accuracy and interpretability of analyses. Ignoring relevant prior information can lead to misleading results, while integrating it within a Bayesian framework offers a more comprehensive and nuanced understanding.

It's important to note that the choice of prior distribution is crucial. An inappropriately chosen prior can bias the results, potentially leading to inaccurate conclusions. Care must be taken to select a prior that reflects the available prior knowledge as accurately as possible, while remaining objective enough to avoid unduly influencing the posterior. Sensitivity analysis, which involves exploring the impact of different priors on the posterior, can be a valuable tool to assess the robustness of the results and understand the influence of prior assumptions. This methodical approach underscores the importance of rigorous planning and critical evaluation in any Bayesian analysis.

The applications of Bayes' Theorem are incredibly vast within bioinformatics. From gene expression analysis to protein structure prediction and phylogenetic inference, Bayesian methods offer a powerful and flexible framework for modeling uncertainty and incorporating prior knowledge. Its iterative nature lends itself well to handling complex biological systems, allowing us to refine our models as we gather more data. The ability to quantify uncertainty, a key feature of Bayesian inference, is particularly important in bioinformatics where biological systems are inherently complex and noisy.

The theorem's power lies not only in its theoretical elegance but also in its practical usability. Numerous software packages and statistical tools are readily available, providing convenient and efficient implementation of Bayesian methods. These tools make it accessible to researchers with varying levels of statistical expertise, fostering broader adoption of Bayesian techniques within the bioinformatics community. This accessibility helps democratize the application of robust statistical approaches across a wide range of bioinformatics problems.

In closing, understanding and applying Bayes' Theorem is fundamental to anyone serious about working with probabilistic modeling in bioinformatics. Its capacity to integrate prior knowledge with new data, and its inherent ability to handle uncertainty, makes it an essential tool in navigating the complexities of biological systems. Mastering this theorem provides a powerful arsenal for unveiling the intricate molecular processes that underpin life. Its flexibility, the iterative refinement of conclusions based on new information, and its ease of implementation makes it an indispensable technique for anyone working to decipher the mysteries of the biological world. Therefore, actively exploring this theoretical framework and practicing its application will undoubtedly enhance the ability to interpret and make meaningful conclusions from biological data. Embrace the power of Bayes; it will enrich your understanding of biological processes.

1.2 Prior and Posterior Distributions

Understanding the prior distribution is crucial; it represents our initial beliefs about a parameter before observing any data. This might stem from previous experiments, theoretical considerations, or even a simple, uninformed guess. For example, if we're estimating the mutation rate in a specific gene, our prior might be based on known mutation rates in similar genes or from broader genomic studies. The key is to choose a prior distribution that reflects our existing knowledge accurately. A poorly chosen prior can skew results, leading to misleading conclusions. Careful consideration of prior information, or its absence, is vital. The flexibility in choosing a prior is a significant advantage of the Bayesian approach, allowing us to incorporate domain expertise effectively.

The posterior distribution, on the other hand, reflects our updated belief after incorporating the observed data. It's the result of combining our prior knowledge with the information gleaned from our experiment, a process guided by Bayes' theorem. Imagine analyzing microarray data to find genes differentially expressed under two conditions. We might start with a prior suggesting similar expression levels in both conditions. However, after analyzing the microarray data, the posterior distribution might show a shift, indicating a clear difference in expression for some genes. This illustrates the power of the Bayesian framework: it allows us to refine our initial beliefs in light of new evidence. Essentially, it allows us to learn from data.

Now, let's delve into practical aspects. Selecting an appropriate prior is paramount. Sometimes, a non-informative prior, representing a lack of strong prior belief, is the most suitable choice. This approach lets the data speak for itself, minimizing the influence of prior assumptions. Other times, an informative prior, reflecting strong prior knowledge, might be more appropriate if substantial prior evidence exists. The choice depends entirely on the context and available information, and a poorly chosen prior can significantly affect the reliability of results. This decision can be a critical part of your analysis and demands cautious consideration.

The practical application often involves computational methods. Many software packages provide tools for Bayesian analysis. These tools often facilitate the selection of prior distributions, calculation of posterior distributions, and the visualization of results. Familiarity with such tools is key. Understanding the output and interpreting the posterior distribution is equally important. For instance, credible intervals, analogous to confidence intervals in frequentist statistics, quantify the uncertainty associated with the estimated parameter. These provide a range of plausible values for the parameter, reflecting the uncertainty inherent in the estimation process.

Another critical aspect is model selection. Bayesian methods offer elegant solutions for model comparison and selection. The Bayes factor compares the posterior probabilities of competing models. In essence, the Bayes factor helps us decide which model is better supported by the data, considering our prior beliefs. This differs substantially from frequentist model selection methods. The Bayes factor provides a quantifiable measure of evidence in favor of one model over another.

Moreover, the ability of Bayesian inference to handle missing data is a huge advantage. In bioinformatics, missing data is prevalent, whether it's due to experimental limitations or inherent biological variability. Bayesian methods can effectively incorporate prior knowledge to estimate missing values, resulting in more robust and complete analyses. This allows us to overcome limitations often encountered in other statistical frameworks.

Finally, let us address limitations. Bayesian inference involves subjective choices, particularly regarding the selection of prior distributions. Subjectivity can introduce biases, albeit often less prominent than the biases in a poorly designed frequentist study. However, transparency is essential; clearly documenting the chosen prior and its rationale is crucial for reproducibility and evaluation by others. Choosing a weakly informative prior and testing sensitivity to prior choices can mitigate these concerns. Also, computational demands can be high, especially for complex models.

In conclusion, understanding and effectively utilizing prior and posterior distributions is central to Bayesian inference in bioinformatics. Mastering this concept unlocks the power of this framework, leading to more robust and informative analyses of biological data. Choosing the right prior, correctly interpreting the posterior, and understanding the computational aspects are critical skills for any bioinformatician employing Bayesian methods.

2 Bayesian Networks in Biology

Bayesian networks offer a compelling framework for modeling complex biological systems, moving beyond simple linear relationships to capture the intricate interplay of multiple factors. Imagine a gene regulatory network; a single gene's expression isn't solely determined by its own promoter region but is profoundly influenced by upstream regulators and downstream targets, often forming feedback loops and complex cascades. A Bayesian network elegantly represents these interactions as a directed acyclic graph, with nodes representing genes or proteins and edges indicating regulatory influences, each edge representing a conditional probability. This probabilistic approach accounts for the inherent uncertainty and noise present in biological data.

Constructing these networks requires sophisticated inference techniques. Methods like the Bayesian Information Criterion (BIC) or the minimum description length (MDL) principle help us choose the most parsimonious network structure that best explains the observed data. These approaches penalize model complexity, preventing overfitting. Alternatively, Markov Chain Monte Carlo (MCMC) methods, such as Gibbs sampling, can explore the vast space of possible network structures, generating samples from the posterior distribution over networks. This offers a rich understanding of the uncertainty associated with the inferred network structure, rather than providing a single "best" network.

Gene regulatory networks (GRNs) are a prime application area. Researchers use gene expression data, often obtained from microarrays or RNA sequencing, to infer the regulatory relationships among genes. The challenge lies in interpreting the complex correlations between gene expression levels and disentangling direct regulatory influences from indirect ones. Bayesian networks provide a powerful tool to address this challenge by explicitly modeling these conditional dependencies, allowing us to uncover hidden regulatory mechanisms. For instance, we could infer that gene A upregulates gene B, which in turn represses gene C, a relationship that might be missed using simpler correlation analyses.

Beyond GRNs, protein-protein interaction (PPI) networks also benefit immensely from a Bayesian approach. High-throughput experiments, like yeast two-hybrid screens or affinity purification coupled with mass spectrometry, generate massive amounts of data on protein interactions. However, these experiments often yield false positives and negatives, introducing considerable uncertainty. Bayesian networks can integrate these noisy datasets with prior knowledge about protein function and localization to construct more reliable PPI networks. By assigning probabilities to interactions and incorporating biological context, we obtain a more robust representation of the cellular network. Further, we can use these networks to predict protein function, discover new drug targets, and study disease mechanisms.

The strength of Bayesian networks in biological systems stems from their ability to incorporate prior knowledge. This is particularly advantageous when dealing with incomplete or noisy datasets. Prior knowledge about known regulatory interactions, protein functions, or pathway annotations can be incorporated as prior probabilities in the network model, guiding the inference process and improving accuracy. For example, we may already know that a specific transcription factor regulates a set of genes involved in a particular metabolic pathway. Including this information in the prior distribution can significantly strengthen our inference and lead to more biologically meaningful results.

Data integration is another key advantage. Often, researchers collect data from multiple sources—gene expression, protein abundance, and genetic interactions. Bayesian networks provide a natural framework for integrating these diverse data types into a unified model, enhancing the robustness and reliability of the resulting network. This integrated approach facilitates a more holistic understanding of biological processes and allows us to test hypotheses about complex interactions spanning different levels of biological organization. For example, we could combine gene expression data with information about protein-DNA binding to construct a more comprehensive model of gene regulation.

While Bayesian networks offer significant advantages, limitations exist. The computational cost of inference can be substantial, particularly for large networks with many nodes and edges. Moreover, the accuracy of the inferred network strongly depends on the quality and quantity of the input data. In cases with sparse or noisy data, the inferred networks may be unreliable. Researchers often employ various strategies to mitigate these challenges, including using efficient algorithms, incorporating prior knowledge, and carefully validating the inferred networks against independent experimental data. Careful consideration of these issues is crucial for successful application of Bayesian networks in biological research. Despite these limitations, their capacity to model complexity, integrate diverse data types, and handle uncertainty make Bayesian networks a vital tool in understanding the intricate workings of biological systems.

2.1 Network Inference Techniques

Several methods exist for inferring biological networks from high-throughput data. These methods leverage the power of Bayesian inference to account for uncertainty inherent in biological systems and noisy data. One common approach utilizes Bayesian networks, directed acyclic graphs where nodes represent variables (e.g., genes, proteins) and edges represent probabilistic dependencies between them. Learning the structure of these networks involves searching through the space of possible graphs, scoring each based on how well it explains the observed data. This process often involves Markov Chain Monte Carlo (MCMC) methods, which allow exploration of the high-dimensional space of network structures by iteratively sampling graphs according to their probabilities. The choice of scoring function is crucial, often balancing the complexity of the network with its ability to fit the data. Strategies like Bayesian Information Criterion (BIC) or Akaike Information Criterion (AIC) help manage the trade-off between model complexity and goodness of fit, preventing overfitting which can lead to inaccurate network representation. For example, in gene regulatory networks, a gene's expression level might depend on the expression levels of its regulators, leading to directed edges from regulators to the regulated gene. Inferring these dependencies from gene expression data, potentially utilizing time-series data for improved accuracy, offers insights into gene regulation mechanisms. Appropriate data preprocessing is paramount here, including normalization and handling of missing values to optimize the accuracy of inference.

Another powerful technique is the use of relevance networks, which leverage correlation or mutual information between variables. High correlations between variables suggest a possible relationship. Bayesian methods refine this by incorporating prior knowledge and quantifying uncertainty associated with the estimated correlations. A challenge, however, lies in distinguishing direct interactions from indirect ones as correlations can emerge even without direct interaction. A potential solution involves incorporating time-series data and other biological knowledge to improve interpretation. Bayesian methods can gracefully integrate such information, enhancing the accuracy of network inference. This highlights the need for careful experimental design, ensuring sufficient sampling across relevant conditions.

Furthermore, Gaussian graphical models (GGMs) offer a valuable approach, particularly suitable when dealing with continuous data such as gene expression levels. GGMs represent conditional dependencies between variables using a precision matrix (the inverse of the covariance matrix), where the absence of an edge between two nodes indicates conditional independence. Bayesian methods, employing prior distributions on the precision matrix parameters, allow for regularization and sparsity enforcement, reducing the number of false positives and improving interpretability. This combats the issue of high dimensionality, a common problem in genomics, where many variables (genes) are measured with relatively few samples. Prior distributions tailored to encourage sparsity, such as those based on the Laplace distribution, contribute significantly to the power of this approach. The Bayesian framework allows for quantifying uncertainty in the inferred network structure, providing confidence measures for each edge. Thus Bayesian inference yields a probability distribution over possible networks rather than a single, definitive answer.

However, challenges remain. Inferring networks from high-throughput data frequently involves dealing with large datasets, computationally intensive algorithms, and high levels of noise. Computational limitations could restrict the scale of the networks that are practically inferable. Additionally, the inherent complexity of biological systems often means that the inferred networks are approximations, potentially missing intricate interactions or highlighting spurious relationships. Careful validation and comparison of results from multiple inference methods are crucial to mitigate these inherent limitations. The accuracy of the inferred network heavily relies on the quality and completeness of the input data. Incomplete data or systematic biases can mislead the inference algorithms, leading to erroneous conclusions. Choosing the right algorithm, and verifying results using established biological knowledge and external validation techniques are always advisable. Ongoing research focuses on developing improved inference algorithms that can efficiently handle larger datasets, incorporate more diverse data types, and account for the dynamic nature of biological networks. The integration of prior knowledge, perhaps from literature or databases, promises improved accuracy and interpretability. The future of network inference will likely involve more sophisticated Bayesian techniques combined with machine learning algorithms.

2.2 Gene Regulatory Networks

Understanding gene regulatory networks (GRNs) is crucial for deciphering the complexities of cellular processes. These networks depict intricate relationships between genes, where the expression of one gene influences the expression of others, creating a cascade of effects that ultimately determine cellular behavior. Visualizing these networks as interconnected nodes and edges, where nodes represent genes and edges represent regulatory relationships (activation or repression), allows for a systematic understanding of gene regulation. This is where Bayesian inference proves invaluable.

Bayesian methods are particularly well-suited for analyzing GRNs because they allow us to incorporate prior knowledge—such as known transcription factor-binding sites or gene ontology annotations—into our model. This prior knowledge can significantly improve the accuracy and robustness of our inferences, especially when dealing with noisy or incomplete datasets, a common challenge in biological research. The incorporation of prior knowledge isn't just a theoretical benefit; it fundamentally shifts the inference from solely data-driven to a hybrid approach, blending empirical evidence with established biological insight. This synergistic approach is far more powerful than either method alone.

Inferring the structure of a GRN involves determining the presence and directionality of regulatory interactions between genes. This is commonly achieved using various computational methods, often involving statistical modeling and machine learning algorithms. One such method might be to model gene expression data using Bayesian networks, which represent the probabilistic dependencies between genes. By learning the parameters of these networks, we can identify likely regulatory interactions. Consider, for instance, a scenario where gene A's expression is consistently high when gene B's expression is low. A Bayesian network could infer a negative regulatory interaction from this correlation, highlighting a potential repressive effect of gene A on gene B.

However, it's critical to acknowledge that the inference process isn't foolproof. The accuracy of inferred GRNs depends heavily on the quality and quantity of the input data, as well as the assumptions underlying the chosen modeling technique. The presence of noise, experimental errors, and confounding factors can all lead to incorrect inferences. Therefore, thorough validation and model evaluation are essential steps in any GRN inference process. Utilizing rigorous cross-validation techniques, perhaps employing Bayesian model averaging to combine results from multiple models, ensures more reliable outcomes. It also emphasizes the importance of integrating data from diverse sources, like those from different experiments, to strengthen the overall inference.

Beyond network inference, Bayesian methods are also valuable for analyzing the dynamics of GRNs. We can use dynamic Bayesian networks to model how the expression levels of genes change over time in response to various stimuli or conditions. This allows us to understand the temporal relationships between genes and predict how the network will behave under different circumstances. The modeling of dynamic systems provides a crucial insight into the complex feedback loops and intricate interactions that are often the defining characteristics of cellular control systems. It opens up avenues for studying diseases and identifying promising drug targets. For example, we might use this dynamic approach to model the effects of a drug on the expression pattern of genes involved in a particular disease pathway. This detailed modeling allows for the generation of targeted and effective therapeutic interventions.

Moreover, Bayesian approaches can help us address the challenges of handling uncertainty and variability inherent in biological systems. Gene regulation is a stochastic process, influenced by various random factors. Bayesian methods explicitly incorporate this uncertainty into the inference process, providing more realistic and nuanced representations of GRNs. The inherent probabilistic nature of the Bayesian framework makes it especially well-suited for handling noisy data and incomplete information. By quantifying uncertainty, we can identify those aspects of the network that are more precisely defined, which can further guide the selection of additional experiments that would clarify aspects of the system still shrouded in uncertainty.

Finally, the power of Bayesian inference in GRN analysis extends to its capacity for hypothesis testing and model comparison. By comparing different GRN models using Bayesian model selection criteria, such as Bayes factors, we can determine which model best explains the observed data. This allows for a more objective evaluation of competing hypotheses about the underlying regulatory mechanisms and further aids in refining the models for increased accuracy. A careful comparison of models allows researchers to move past anecdotal observations toward a more rigorous understanding of biological regulation.

In summary, Bayesian inference provides a powerful and versatile toolkit for investigating gene regulatory networks. By integrating prior knowledge, handling uncertainty, and enabling model comparison, Bayesian methods offer a rigorous and informative approach to unraveling the complexities of gene regulation and its essential role in cellular function. The ability to handle noisy biological data, quantify uncertainty and incorporate prior knowledge offers a significant advantage over alternative techniques. This makes Bayesian inference a powerful tool for anyone seeking to understand this critical area of biological study.

2.3 Protein Interaction Networks

Understanding protein interactions is crucial for deciphering the complexities of cellular processes. These interactions, often transient and dynamic, form intricate networks that govern everything from signal transduction to metabolic regulation. Visualizing these networks as graphs, where proteins are nodes and interactions are edges, allows us to appreciate their scale and complexity, revealing emergent properties not evident at the individual protein level. The sheer volume of data generated by high-throughput techniques like yeast two-hybrid screens and affinity purification coupled with mass spectrometry presents a significant computational challenge. However, Bayesian methods provide a robust framework for analyzing this data, inferring the most likely network structure given the observed interactions.

Bayesian approaches offer distinct advantages over traditional methods. Unlike frequentist techniques, which focus solely on the observed data, Bayesian inference incorporates prior knowledge, which can be particularly valuable in the context of protein interaction networks. We might possess prior information about known interactions from previous research, or we could utilize a prior that reflects our understanding of the cellular context, such as the likelihood of interactions between proteins located within the same cellular compartment. Integrating this information helps refine our estimates of interaction probabilities, leading to more accurate and reliable network reconstructions. This is especially useful when dealing with noisy or incomplete data, a common issue in high-throughput experiments.

The key to effectively applying Bayesian methods lies in selecting the appropriate model. Several models exist, each with its strengths and limitations. For instance, we can use simple probabilistic models assuming independence between interactions, or we can choose more complex models capturing dependencies between proteins. The choice depends on the specific biological question, the available data, and the computational resources. A simple model might suffice for a preliminary exploration of a small-scale network, while a sophisticated model would be necessary to analyze large-scale interaction maps. Model complexity is often balanced against the computational cost of inference.

Inferring the structure of a protein interaction network involves estimating the probability of an interaction between each pair of proteins. This estimation relies on the available data, which typically includes a combination of positive and negative evidence. Positive evidence arises from experimental observations of direct interactions, while negative evidence represents the absence of detected interactions. However, the absence of evidence doesn't equate to the evidence of absence. Bayesian methods account for this uncertainty, providing probability distributions over interaction states rather than simply assigning binary classifications (interaction or no interaction).

The posterior distribution, which summarizes our updated belief after observing the data, plays a central role in the analysis. This distribution quantifies the probability of different network structures, given the data and the prior. From this distribution, we can identify the most probable network structure, as well as the probability of each individual interaction. This provides a probabilistic framework for evaluating the confidence in our network reconstruction. Furthermore, the posterior distribution reveals relationships between proteins, even in the absence of direct interaction data. For instance, if two proteins frequently appear in the same complexes, a Bayesian model can highlight the statistically significant co-occurrence as indirect evidence for interaction, a powerful tool for detecting functional modules.

Analyzing the inferred networks requires careful consideration of several factors. Network properties like degree distribution (the number of interactions per protein) or clustering coefficient (the density of interactions within a neighborhood) can provide insights into the overall network topology and the functional roles of individual proteins. Hub proteins, those with a high number of interactions, frequently play crucial roles in cellular processes and are of particular interest. Identifying such hubs within inferred networks can direct further experimental investigations. However, it's crucial to remember that the inferred networks are probabilistic representations of reality, subject to uncertainty. The associated confidence scores should always be considered when interpreting the results. Robustness analysis using various prior models and parameters can evaluate the sensitivity of the inferred networks to uncertainties in the experimental data.

To integrate these techniques effectively into a practical workflow, one needs a solid understanding of Bayesian inference principles and the available computational tools. Several software packages offer specialized functionalities for Bayesian network inference from protein interaction data. Familiarity with these tools and their limitations is paramount. However, technical proficiency is only part of the equation. Biological context and proper data pre-processing are crucial for meaningful results. Carefully selecting and pre-processing interaction data, removing false positives and dealing with missing data, is essential for reliable inferences. The interpretation of the resulting network should be closely connected to relevant biological knowledge. The resulting network should not be viewed in isolation; it needs to be compared to existing knowledge and integrated into the broader biological context. This process often involves iteratively refining the model based on new evidence and biological insight. Bayesian inference is not merely a computational tool; it's a framework for iterative learning and hypothesis refinement, central to modern biological discovery.

3 Hidden Markov Models in Bioinformatics

Hidden Markov Models (HMMs) are invaluable tools in bioinformatics, particularly adept at handling sequential data. Imagine a protein sequence: a string of amino acids, each with its own properties. An HMM can model this sequence as a series of hidden states—for example, alpha-helices, beta-sheets, or random coils—that generate the observed amino acid sequence. The beauty of HMMs lies in their ability to infer these hidden states, even without directly observing them, providing insights into the protein's three-dimensional structure.

Unlike simpler methods, HMMs incorporate uncertainty inherent in biological data. They don't just give you a single, definitive answer; instead, they provide probability distributions, indicating the likelihood of different states at each position in the sequence. This probabilistic framework is crucial because biological processes are often noisy and complex, and HMMs elegantly account for this inherent variability. This is especially relevant in applications like gene prediction, where the boundaries between coding and non-coding regions are often ambiguous.

Gene prediction is a classic example of HMM applications. Consider a genome: a long sequence of DNA bases. Genes are embedded within this sequence, but identifying them isn't straightforward. Using an HMM, we can model the gene structure as a series of hidden states: promoters, exons (coding regions), introns (non-coding regions), and terminators. Each state emits a particular sequence of DNA bases, with probabilities that reflect the statistical properties of real genes. By training an HMM on known gene sequences, we can effectively "teach" it to identify potential genes in new sequences, giving a probability of a specific region being a gene. This is particularly powerful as many genes share the same basic structure of exons and introns, but their sequences might differ.

Phylogenetic inference, the reconstruction of evolutionary relationships between species, also benefits enormously from HMMs. Consider aligning multiple DNA sequences from different organisms. Variations arise due to mutations over time. An HMM can model the evolutionary process as a series of hidden states representing ancestral sequences, with transitions between states reflecting evolutionary events such as insertions, deletions, or substitutions. By considering the probabilities of such events, we can construct phylogenetic trees that reflect the evolutionary history of the species, and not just the order in which substitutions occurred.

HMMs are also exceptionally powerful in the field of motif finding. Biological sequences often contain conserved short patterns called motifs, which indicate functional elements. For instance, in DNA sequences, promoters often contain specific motifs that regulate gene expression. HMMs allow us to model the presence or absence of these motifs within a longer sequence. The algorithm considers the probability of a motif occurring in various regions of the sequence and provides a probability of a given region containing a conserved motif, even if the motif sequence shows some variation.

Comparative genomics, the study of multiple genomes to understand evolutionary relationships and functional elements, also leverages HMMs effec-

tively. By comparing sequences across multiple species, we can detect conserved regions using HMMs, which may indicate the presence of functional elements. In contrast to simpler sequence comparison methods, HMMs offer a sophisticated way to address evolutionary changes that might have obscured the similarity between functional elements in different species.
Beyond these core applications, HMMs are continually being refined and adapted to tackle new bioinformatics challenges. For instance, probabilistic extensions are being developed to incorporate additional factors like epigenetic modifications that are crucial for a holistic understanding of the genome, and not just its sequence. As such, despite their relative simplicity, HMMs remain an integral part of bioinformatician's toolbox, providing powerful and flexible methods for analyzing and modeling biological data. Their probabilistic nature makes them especially well-suited for the inherently uncertain world of biological processes, and their capability to infer hidden structures within sequence data renders them adaptable to various challenges. The ongoing development of novel algorithms and extensions ensures that HMMs will continue to play a vital role in advancing our understanding of the complex intricacies of biological systems. Furthermore, their relatively low computational demands, in comparison to some other Bayesian methods, contribute to their widespread applicability, even with massive datasets. The ease with which they can be applied and interpreted is perhaps the biggest reason for their lasting success.

3.1 Sequence Alignment

Sequence alignment is a fundamental task in bioinformatics, crucial for comparing biological sequences like DNA, RNA, or protein sequences. Its purpose is to identify regions of similarity, which often indicate functional, structural, or evolutionary relationships. We use algorithms to arrange the sequences optimally, highlighting conserved regions and introducing gaps where necessary to maximize the alignment score. This score reflects the similarity between the sequences, helping researchers infer evolutionary relationships or predict protein function. Different scoring systems exist, considering factors such as the types of residues involved, the cost of introducing gaps, and the length of the alignment.

The simplest alignment methods employ pairwise comparison, evaluating similarity between only two sequences at a time. Dynamic programming, a powerful algorithmic approach, offers an effective way to compute optimal alignments, guaranteeing the best possible alignment according to a chosen scoring scheme. However, as the number of sequences increases, the computational cost of pairwise dynamic programming grows exponentially, making it impractical for large datasets. Therefore, more sophisticated approaches are necessary for multiple sequence alignments.

Multiple sequence alignment (MSA) extends pairwise alignment to three or more sequences simultaneously, revealing conserved regions that might be missed in pairwise comparisons. Progressive alignment, a commonly used heuristic method, constructs MSAs incrementally, starting with a pair of most similar sequences, then gradually adding others based on their similarity to existing alignments. This approach is computationally efficient, but it can be sensitive to the initial alignment choices, which might affect the accuracy of the final MSA. Iterative refinement methods address some limitations of progressive methods by iteratively improving the alignment, potentially providing more accurate results, though at increased computational cost.

Hidden Markov Models (HMMs) provide a powerful probabilistic framework for sequence alignment. Unlike deterministic methods, HMMs incorporate uncertainty explicitly, modeling both observed sequence data and hidden states that represent underlying biological features. For example, in protein sequence alignment, hidden states might represent conserved regions, variable regions, or insertion/deletion events. This probabilistic approach allows for more robust and statistically well-founded alignments, particularly when dealing with noisy data or sequences with significant divergence. HMMs offer an elegant solution for incorporating prior knowledge, and they're particularly suitable for problems involving alignment and annotation simultaneously, such as gene finding or motif discovery.

The Bayesian framework enhances HMMs further. Instead of relying solely on point estimates of parameters, the Bayesian approach generates posterior probability distributions over model parameters and hidden states. This provides a richer picture of uncertainty, facilitating more informed decision-making. For example, rather than obtaining a single optimal alignment, Bayesian methods can estimate the probability distribution of multiple possible alignments. This ability to quantify uncertainty is valuable, especially in biological applications where data is often incomplete or ambiguous.

Applying Bayesian methods to sequence alignment involves defining a prior distribution over the parameters of the HMM (e.g., emission probabilities and transition probabilities), which could reflect prior biological knowledge. Subsequently, through integration of the likelihood function (probability of observing the data given the model) with the prior distribution, we obtain the posterior distribution. This posterior reflects our updated knowledge after observing the sequence data. Inference from this posterior distribution may involve techniques like Markov Chain Monte Carlo (MCMC) sampling to sample parameters from the complex posterior distributions. These methods make it possible to estimate and evaluate alignment reliability.

Practical applications of Bayesian sequence alignment extend across diverse bioinformatics domains. For example, in phylogenetic studies, aligning multiple homologous DNA sequences informs the construction of phylogenetic trees, revealing evolutionary relationships between species. Similarly, in protein structure prediction, sequence alignments facilitate homology modeling—building structural models based on known structures of similar proteins. In gene finding, accurate sequence alignment helps in identifying protein-coding genes within genomic sequences.

Choosing the appropriate alignment method depends on several factors. The number of sequences, the length of sequences, and the degree of similarity all play vital roles. For only a few sequences, pairwise dynamic programming is often a suitable choice. However, for many sequences, heuristic methods such as progressive alignment are preferred due to their computational efficiency. Bayesian methods are valuable when accounting for uncertainty and incorporating prior knowledge is essential, even if they involve higher computational burden. Regardless of your method, carefully choosing appropriate scoring parameters and critically assessing alignment results, using multiple methods if necessary, remains crucial. The quest for improved alignment accuracy continues to drive research in the field, with ongoing advancements in algorithmic and probabilistic methods leading to improved interpretations of biological sequences. Understanding these different methods allows bioinformaticians to select the most appropriate tool to address specific research questions effectively.

3.2 Gene Prediction

Gene prediction, a cornerstone of bioinformatics, relies heavily on the power of Bayesian inference. We don't simply guess at gene locations; instead, we leverage probabilistic models to refine our understanding based on available data. This data might include DNA sequence information, codon usage biases, or even comparative genomic data from related species. The Bayesian framework allows us to incorporate prior knowledge – perhaps about the typical length of genes or the presence of specific promoter sequences – to construct a more robust and accurate prediction model. The inherent uncertainty in biological data is naturally accommodated within this framework, providing more reliable predictions than those based on deterministic approaches. Ultimately, Bayesian inference enables us to move beyond simple heuristics and make informed probabilistic statements about the presence, location, and even function of genes.

Unlike simpler methods, Bayesian gene prediction accounts for the inherent ambiguity present in genomic sequences. Consider the challenge of identifying coding regions amidst a sea of non-coding DNA. Simple algorithms might misinterpret stretches of random sequence as genes. However, a Bayesian approach allows us to assign probabilities to different potential gene models. We might weigh the evidence supporting a potential gene against the prior probability of a random occurrence, resulting in a more refined and realistic probability of the gene's existence. This probabilistic approach allows us to quantify the uncertainty associated with each prediction, providing a richer and more complete picture than a simple "yes" or "no" answer.

The power of Bayesian inference is further amplified when combined with Hidden Markov Models (HMMs). HMMs are particularly well-suited to modeling the sequential nature of genomic data. They represent the gene prediction problem as a hidden Markov process, where the "hidden" states correspond to different regions within a gene (e.g., exons, introns). The observable states are the DNA sequences themselves. By learning the transition probabilities between these hidden states and the emission probabilities of observing certain sequences given a particular state, we can create a powerful model for gene prediction. The Bayesian framework allows us to incorporate prior knowledge into the model parameters, improving prediction accuracy and robustness.

Moreover, integrating comparative genomic data further enhances the power of Bayesian gene prediction. By comparing the genome of interest to the genomes of related species, we can identify conserved regions likely to represent functional genes. The Bayesian approach naturally incorporates this information by adjusting the prior probabilities of gene existence based on the observed conservation patterns. This makes the prediction process more reliable and helps avoid spurious gene predictions that might arise from chance similarities in unrelated sequences. A truly Bayesian framework will not only account for this comparative data, but also assess its reliability; that is, some genomes are more closely related than others and should have greater influence on the predictions. Therefore, the weights and influences of this comparative data must themselves be subject to Bayesian reasoning, leading to more nuanced and reliable gene predictions.

In practice, the implementation of Bayesian gene prediction often involves sophisticated computational techniques, such as Markov Chain Monte Carlo (MCMC) methods. MCMC algorithms allow us to sample from the posterior distribution of the gene model parameters, providing a comprehensive representation of the uncertainty inherent in the predictions. These methods are computationally intensive, but the payoff is significant – we obtain a far more accurate and nuanced understanding of gene structure and location than traditional deterministic approaches. For example, posterior predictive checks, an integral part of Bayesian modeling, help assess the quality of our model and flag potential problems. They allow us to validate the model's ability to reproduce observed patterns in the data. This thoroughness makes Bayesian methods less prone to the pitfalls that other methods may fall into.

Furthermore, the Bayesian framework is flexible and adaptable, enabling us to incorporate diverse types of genomic data. We can easily extend the model to include information from RNA-Seq data, which provides direct evidence of gene expression. This integration greatly strengthens the accuracy of gene prediction, allowing us to distinguish between actively transcribed genes and pseudogenes. The flexibility and power of the Bayesian approach are particularly important in analyzing the increasingly complex genomic data generated by modern high-throughput sequencing technologies. It offers a principled way to integrate the vast amount of information now available, and allows us to gain more complete insights into the intricate world of gene regulation and function. The ability to naturally combine diverse sources of evidence makes it a powerful and highly adaptable method for addressing the challenges of gene prediction in the modern era of biological research.

Finally, the practical application of Bayesian gene prediction extends far beyond simple gene identification. By leveraging the probabilistic nature of the model, we can obtain valuable insights into the regulatory mechanisms controlling gene expression. We might identify potential promoter regions or transcription factor binding sites by analyzing the posterior distribution of the model parameters. This is invaluable for understanding the complex interplay of genetic factors that influence gene activity. This makes it a robust method to analyze and interpret complex datasets, resulting in a far more comprehensive biological understanding. The ultimate goal is not just to find genes, but to understand their function and how they contribute to the complex tapestry of life. Bayesian inference offers a powerful pathway to achieve this goal.

3.3 Phylogenetic Inference

Constructing a phylogenetic tree, a visual representation of evolutionary relationships, is a fundamental task in bioinformatics. We utilize Bayesian methods to infer these trees because they elegantly handle uncertainty inherent in biological data, providing not just a single "best" tree, but a probability distribution over all possible tree structures. This allows us to quantify our confidence in different branches of the tree and identify those relationships supported by stronger evidence. Unlike simpler methods, Bayesian inference incorporates prior knowledge about evolutionary processes, making it particularly powerful when dealing with limited or ambiguous data.

The process begins with selecting an appropriate evolutionary model. This model describes how DNA or protein sequences change over time, accounting for factors like substitution rates and the probability of different types of mutations. Choosing the right model is critical; a poorly chosen model can lead to inaccurate or misleading results. Model selection often involves comparing multiple models using metrics such as the Bayes factor or Deviance Information Criterion (DIC), assessing how well each model explains the observed data. Careful consideration should be given to the biological system under study; for example, fast-evolving sequences might require a model that accommodates higher mutation rates than slowly evolving ones.

Once a model is selected, the Bayesian phylogenetic inference algorithm proceeds through a Markov Chain Monte Carlo (MCMC) approach. MCMC methods simulate a random walk through the space of possible phylogenetic trees, gradually converging on a distribution that reflects the posterior probability of each tree given the data and the chosen model. In simpler terms, the algorithm explores numerous tree topologies and branch lengths, weighting each configuration according to its likelihood of generating the observed sequences. This iterative process is computationally intensive, necessitating the use of specialized software. Popular packages include MrBayes and BEAST, offering various options for model selection and parameter settings.

Monitoring the MCMC run is vital to ensure convergence and proper sampling. Convergence diagnostics, such as trace plots and effective sample sizes (ESS), should be carefully examined. Trace plots visualize the evolution of parameters during the MCMC run, indicating whether the algorithm has reached a stable state. Low ESS values may suggest insufficient sampling, necessitating a longer run or adjustment of MCMC settings. A thorough analysis of these diagnostics is essential to ensure the reliability of the resulting phylogenetic tree and its associated probabilities.

Interpreting the resulting phylogenetic tree requires understanding the concept of posterior probabilities. These probabilities represent the confidence level associated with each branch of the tree, reflecting the strength of evidence supporting its existence. High posterior probabilities, typically above 0.95, indicate strong support for a particular clade (a group of organisms sharing a common ancestor). However, low probabilities warn against overinterpreting those relationships, suggesting further investigation might be needed to clarify evolutionary relationships. Focusing solely on the highest probability tree often neglects the inherent uncertainty in phylogenetic reconstruction; the entire posterior distribution should be considered when drawing conclusions.

One powerful advantage of Bayesian inference is its capacity to incorporate prior information into the analysis. Prior distributions reflect our pre-existing beliefs or knowledge about evolutionary relationships, which might come from previous studies or established taxonomies. Carefully designed prior distributions can be highly informative, particularly when dealing with sparsely sampled datasets. However, overly informative priors can bias the results, overriding the data's signal. Therefore, it is crucial to carefully construct and evaluate prior distributions to strike a balance between incorporating valuable prior knowledge and avoiding unwarranted influence on the posterior. This process requires domain expertise and a good understanding of the limitations of any prior assumptions.

Furthermore, Bayesian approaches provide a framework for comparing different phylogenetic models and assessing the uncertainty inherent in the inference process. By comparing models using the Bayes factor, we can assess which model best explains the data. In the context of phylogenetic inference, this allows for investigating the impact of different evolutionary models on the resulting tree topology and branch lengths. This is crucial because different models can lead to substantially different evolutionary inferences. The Bayesian framework accommodates this uncertainty systematically, providing a more robust and realistic assessment of evolutionary history.

The application of Bayesian methods in phylogenetic inference significantly enhances our ability to reconstruct the evolutionary history of biological sequences. By quantifying uncertainty, accommodating prior information and providing tools to compare models, Bayesian approaches provide a powerful and versatile framework to generate more robust and reliable phylogenetic trees. As datasets grow and computational resources improve, Bayesian methods will only become more central to evolutionary biology and bioinformatics research. The accurate interpretation of phylogenetic trees derived through this methodology depends on careful consideration of several factors, including model selection, MCMC convergence diagnostics, and the careful interpretation of posterior probabilities. Understanding these aspects is essential to harness the full potential of Bayesian phylogenetic inference.

3.4 Motif Finding

Finding motifs within biological sequences is crucial for understanding gene regulation and other fundamental biological processes. These motifs, often short sequences of DNA or RNA, act as binding sites for proteins, influencing gene expression. Identifying them directly from raw sequence data can be challenging due to noise and the inherent variability in motif instances. Bayesian methods offer a robust framework to address these challenges.

Unlike simpler approaches that rely on stringent thresholds, Bayesian methods allow for probabilistic incorporation of prior knowledge and uncertainty.

For instance, we might have prior knowledge about the consensus sequence of a particular motif family or information about the position weight matrix (PWM) derived from previous experiments. A Bayesian approach seamlessly integrates such information into the motif-finding process, leading to more accurate and reliable predictions, even with limited data.

The core of Bayesian motif finding often involves the use of Hidden Markov Models (HMMs). An HMM models the sequence as a series of hidden states, representing the presence or absence of the motif, and observable states, representing the individual nucleotides. The model parameters, such as the emission probabilities (the probability of observing a particular nucleotide given a hidden state) and transition probabilities (the probability of transitioning between hidden states), are learned from the data using Bayesian inference techniques. This iterative process refines our understanding of the underlying motif structure, gradually improving the model's accuracy in identifying true motifs while mitigating false positives.

A key advantage of this Bayesian approach lies in its ability to quantify uncertainty. Rather than simply declaring a sequence as a motif or not, the Bayesian framework provides a probability that a given sequence represents a true motif. This probability takes into account both the observed data and our prior knowledge, offering a nuanced perspective that avoids the binary nature of many non-Bayesian methods. This probability provides crucial insights into the reliability of predictions, thereby aiding in the interpretation of results and in the subsequent biological hypothesis generation.

Consider the task of identifying transcription factor binding sites. These sites often exhibit sequence variability; a perfect match to a consensus sequence is uncommon. Traditional methods might miss such imperfect but functionally relevant sites. However, a Bayesian approach can incorporate this sequence variability into the model. This is achieved by modeling the emission probabilities as distributions rather than fixed values, allowing for deviations from the consensus sequence without penalizing true motif instances. This flexibility improves sensitivity, capturing a broader range of biologically meaningful sequences.

Furthermore, Bayesian methods readily handle multiple motifs within a single sequence. Many sequences contain overlapping or nearby occurrences of different motifs, leading to complexities that simpler algorithms often struggle to resolve. Bayesian models can be extended to simultaneously model multiple motifs, inferring their positions and frequencies within the sequence. This capability significantly enhances the applicability of the method for tasks such as analyzing promoter regions, which often contain many regulatory elements.

One common method is Gibbs sampling, a Markov Chain Monte Carlo (MCMC) technique. This involves repeatedly sampling motif instances from the sequence, updating the model parameters based on the sampled motifs, and iterating until convergence. The final output doesn't just give you a single "best" motif; instead, it provides a distribution of possible motifs, reflecting the inherent uncertainty in the data and allowing for a more comprehensive understanding.

The choice of prior distributions in a Bayesian motif-finding model significantly impacts the results. Informative priors, based on existing knowledge about similar motifs, can improve accuracy, particularly when data is scarce. However, overly strong priors can also bias the results. The careful selection of prior distributions is a critical step in developing a successful Bayesian motif-finding model. Often, a sensitivity analysis examining the effect of different prior choices on the results is warranted.

Beyond Gibbs sampling, other Bayesian techniques like variational inference can be used. Variational inference provides a deterministic approximation to the posterior distribution, offering an alternative to MCMC that can be computationally more efficient for very large datasets. However, this efficiency sometimes comes at the cost of a less accurate representation of the true posterior. The choice between these approaches depends on the specific dataset and computational resources available. Computational efficiency is particularly important when dealing with genome-scale datasets.

In conclusion, Bayesian motif finding offers a powerful and versatile framework for uncovering hidden regulatory sequences in biological data. By incorporating prior knowledge and accounting for uncertainty, Bayesian methods provide more accurate, reliable, and biologically interpretable results compared to many alternative approaches. The probabilistic nature of these methods not only delivers motif locations but also offers confidence estimates, enriching our understanding of these fundamental molecular processes. The flexibility to handle multiple motifs and varying levels of sequence similarity makes these methods particularly suited to the complexities of biological data analysis.

3.5 Comparative Genomics

Comparative genomics leverages the power of Bayesian inference to uncover evolutionary relationships and functional insights across multiple genomes. Unlike traditional methods focusing solely on sequence similarity, Bayesian approaches incorporate prior knowledge and uncertainty, leading to more robust and informative analyses. This allows us to move beyond simple pairwise comparisons, generating a much richer understanding of the evolutionary processes shaping genomes.

For beginners, a fundamental step is understanding phylogenetic inference using Bayesian methods. These methods provide a probabilistic framework for constructing evolutionary trees, representing the relationships between different species or strains. Unlike maximum likelihood approaches, which only provide point estimates, Bayesian inference yields probability distributions over tree topologies, branch lengths, and other parameters. This allows researchers to quantify the uncertainty associated with their phylogenetic reconstructions and assess the support for different evolutionary scenarios. Software packages like MrBayes and BEAST provide readily available tools to implement these analyses. Familiarity with these tools, alongside an understanding of the underlying Bayesian methodology, is crucial.

Next, consider the application of Bayesian methods in identifying orthologous genes. Orthologous genes are those that share a common ancestor and are often involved in similar biological functions. Determining orthology is key to understanding gene evolution and functional conservation across species. Traditional methods rely heavily on sequence similarity alone; however, using Bayesian methods, we can incorporate evolutionary models and prior information about gene families and functional annotations to improve orthology prediction accuracy. The result is a more refined understanding of how genes evolve and how their functions are conserved. Integrating data from multiple genomes can further refine this process, resulting in a more comprehensive view of the entire evolutionary process.

Furthermore, exploring genomic rearrangements offers significant insights into genome evolution. Bayesian methods can model these rearrangements, which involve events like inversions, translocations, and duplications. By considering the probabilities of different rearrangement scenarios, we can reconstruct the evolutionary history of genomes and infer the underlying mechanisms driving genomic change. This isn't just about recognizing the rearrangements; Bayesian models help us estimate the rates of these events and determine their impact on gene order and function. For example, Bayesian approaches can pinpoint the most likely evolutionary trajectory that produced the current genome arrangement from an ancestral state. This provides a deeper understanding of evolutionary changes than simply listing genomic rearrangements.

Finally, we can move beyond individual genes and gene families, analyzing entire genomes within a comparative context using Bayesian methods. This involves modeling the evolution of genomic features such as GC content, codon usage bias, or repetitive element density. By integrating these diverse data types, we can gain a comprehensive view of the complex interplay between different genomic characteristics. For instance, comparing these features across many genomes enables testing hypotheses concerning the influence of factors such as selection pressures or mutation rates on the genomic architecture of different organisms. A Bayesian approach offers a robust framework to synthesize this complex data and construct integrated models of genome evolution. By combining multiple data types and incorporating prior knowledge, we can construct more accurate and comprehensive models of the evolutionary forces shaping the genomes we observe. This offers a far more nuanced understanding of evolution than simpler comparison methods. Such holistic analysis would be challenging with methods that lack the Bayesian ability to handle complex interdependencies. Remember, each of these steps builds on the previous one, culminating in a robust comparative genomic study.

4 Bayesian Methods for Microarray Data

Microarray data, with its inherent noise and complexity, presents a unique challenge for biological inference. Bayesian methods offer a powerful and flexible framework for addressing these challenges, allowing for the incorporation of prior knowledge and the quantification of uncertainty. Unlike frequentist approaches that focus solely on point estimates, Bayesian methods provide a full posterior distribution, offering a richer and more nuanced understanding of the data. This allows researchers to not only identify differentially expressed genes but also quantify the confidence associated with these findings. This comprehensive approach is crucial when dealing with the inherent variability in microarray experiments.
A common application of Bayesian methods in microarray analysis is differential gene expression analysis. Traditional methods often rely on stringent thresholds, potentially leading to both false positives and false negatives. Bayesian approaches, however, provide a more sophisticated way to identify genes with altered expression levels across different conditions. By modeling the expression levels as probability distributions, we can quantify the probability that a gene is differentially expressed, taking into account both the observed data and prior beliefs about gene expression patterns. This proba-

bilistic approach allows for a more nuanced interpretation of the results, providing confidence intervals and posterior probabilities that offer more reliable insights.

Furthermore, Bayesian methods excel at handling missing data, a common problem in microarray experiments. Unlike frequentist methods that often require imputation or exclusion of incomplete data points, Bayesian methods can naturally integrate missing values into the analysis. This is achieved through the incorporation of prior distributions that reflect our beliefs about the missing data, enabling a more robust analysis that makes the most efficient use of all available information. The resultant analysis is therefore more complete and less susceptible to bias introduced by data imputation methods.

Beyond differential expression analysis, Bayesian methods are equally valuable in clustering and classification tasks. These methods are especially useful for exploring relationships between genes and samples, creating an easier and more informative pathway for researchers to analyze their data. Bayesian clustering approaches, for example, allow for the probabilistic assignment of genes or samples to different clusters based on their expression profiles. This approach avoids the rigid assignment of traditional clustering algorithms, providing a more flexible and less arbitrary means to identify gene groups and sample classes. This allows scientists to discover underlying biological relationships in their datasets with greater confidence.

The incorporation of prior information is a cornerstone of Bayesian methods and provides a significant advantage in microarray analysis. Researchers can leverage prior knowledge from literature, databases or previous experiments to inform their analysis. This might include information about gene function, pathway membership or expression patterns in similar studies. By incorporating this prior information into the Bayesian model, we can improve the accuracy and precision of our inferences, particularly when dealing with limited

data or noisy measurements. This crucial aspect is critical for robust analyses, particularly for the more challenging biological problems.
The application of Markov Chain Monte Carlo (MCMC) methods is essential for the practical implementation of Bayesian methods in microarray analysis. MCMC algorithms, such as Gibbs sampling or Metropolis-Hastings, allow for the estimation of posterior distributions even for complex models. However, the computational burden associated with MCMC can be substantial, especially when dealing with large datasets. Therefore, carefully selecting efficient MCMC algorithms and making use of computational resources are key aspects of implementing Bayesian methodologies effectively. Improvements in computational power are allowing researchers to analyze larger datasets and more complex models, leading to advancements in this area of study.
Finally, the interpretation of Bayesian results requires careful consideration of both the posterior distributions and the associated uncertainties. Visualizing posterior distributions can aid in identifying differentially expressed genes and assessing the degree of uncertainty associated with these findings. Summary statistics, such as posterior means and credible intervals, provide valuable information for decision-making, aiding researchers in drawing conclusions based on the evidence from their analysis. Moreover, understanding the limitations of the model and the assumptions made are crucial for proper interpretation and reporting of the findings, ensuring responsible use of this powerful analytical method. The increasing availability of user-friendly software packages further simplifies the implementation and interpretation of Bayesian methods, making this approach increasingly accessible to researchers across disciplines.

4.1 Differential Expression Analysis

Analyzing gene expression data using a Bayesian framework offers several advantages over traditional frequentist methods. It allows for the incorporation of prior knowledge about gene expression levels, which can be particularly useful when dealing with limited sample sizes or noisy data. For instance, if we have prior information about the expression levels of certain genes under specific conditions from previous studies, we can incorporate this information into our analysis to obtain more accurate and robust results. This prior knowledge might be represented as informative prior distributions in the Bayesian model, improving estimates.

The core of Bayesian differential expression analysis lies in modeling the expression levels of genes across different conditions. We typically model the expression data as coming from a probability distribution, such as the negative binomial distribution, often appropriate for count data like gene expression measurements obtained from RNA sequencing. This distribution accounts for both the mean expression level and the variability in expression observed across replicates. Parameters of this distribution, such as the mean expression level for each gene under each experimental condition, are then treated as random variables.

Within the Bayesian framework, we assign prior probability distributions to these parameters. These priors can be uninformative (e.g., weak priors reflecting minimal prior knowledge) or informative (incorporating prior knowledge as discussed earlier). The specific choice of prior will depend on the available prior information and the researcher's beliefs. After observing the gene expression data, we use Bayes' theorem to update our beliefs about these parameters, resulting in posterior distributions. The posterior distributions summarize our knowledge about the gene expression levels after incorporating the data, reflecting both the observed data and the prior information.

Determining which genes show differential expression is then done by comparing the posterior distributions of the mean expression levels under different conditions. We can calculate the probability that a given gene is differentially expressed, which is often based on whether there's substantial overlap between the posterior distributions. A commonly used method involves calculating the Bayes factor, comparing the posterior odds of differential expression against the prior odds. A Bayes factor above a certain threshold indicates substantial evidence for differential expression. This approach offers a more nuanced assessment of differential expression compared to simple p-values used in frequentist methods.

Software packages implementing Bayesian methods for differential expression analysis exist, making the process relatively accessible. These packages often provide various functionalities, allowing users to specify different prior distributions, assess model fit using posterior predictive checks, and visualize the results. They typically provide comprehensive outputs including posterior distributions, credible intervals, and probabilities of differential expression for each gene. This facilitates the interpretation of results and improves upon the decision-making process, leading to more conclusive inferences.

Furthermore, the Bayesian approach provides a natural framework for incorporating additional information into the analysis. For instance, we might incorporate information about gene ontology annotations, known biological pathways, or prior knowledge about gene regulation. This can further improve the accuracy and interpretation of the differential expression results. By incorporating these data types the statistical power of the analysis increases, and biologically more meaningful results are generated. Such integrative approaches offer a more complete understanding of the biological processes underlying the observed gene expression changes.

One potential hurdle for beginners is the computational intensity of Bayesian methods. Evaluating posterior distributions often requires sophisticated computational techniques such as Markov Chain Monte Carlo (MCMC) methods. However, the computational burden is being constantly mitigated by advancements in algorithms and hardware. Many readily available software packages cleverly handle these computations, minimizing the technical challenges for users. Therefore, while an understanding of the underlying computational methods is helpful, it is not strictly necessary for applying Bayesian methods for differential expression analysis. Focusing on the interpretation of results remains crucial.

In summary, adopting a Bayesian approach for differential expression analysis provides a robust and flexible way to extract meaningful information from high-throughput gene expression data. This approach allows the incorporation of prior knowledge, enabling more accurate and reliable inferences, particularly when dealing with limited sample sizes or noisy data. While initially the computational aspects might seem daunting, readily available software packages have simplified this aspect, empowering researchers to focus on the interpretation and biological implications of their results. The transition from frequentist techniques to Bayesian methodology is a powerful step towards a more comprehensive and nuanced understanding of molecular mechanisms.

4.2 Clustering and Classification

Microarray data, rich with gene expression levels, often presents a challenge: how to organize the vast amount of information into meaningful groups. Bayesian clustering tackles this by probabilistically assigning genes to clusters based on their expression profiles. Unlike traditional clustering methods that offer a single, definitive grouping, Bayesian clustering provides a probability distribution over possible cluster assignments, reflecting the uncertainty inherent in the data. This approach is particularly useful when dealing with noisy or ambiguous data, a common occurrence in high-throughput experiments. The flexibility of Bayesian methods allows for the incorporation of prior knowledge, such as known gene functions or pathways, which can improve clustering accuracy and provide biologically meaningful interpretations. We can use Dirichlet Process Mixture Models, which are a powerful tool for clustering data when the number of clusters is unknown. They allow the model to automatically determine the appropriate number of clusters based on the data, avoiding the need for arbitrary pre-specification, a significant advantage over many frequentist methods. Think of it like this: you're sorting marbles of varying colors and sizes. A traditional method might force you to sort into a pre-defined number of containers, possibly leading to unnatural groupings. A Bayesian approach, however, allows the number of containers to emerge naturally from the characteristics of the marbles themselves, resulting in more intuitive and accurate groupings.

Bayesian classification, on the other hand, goes a step further. It aims to assign each gene to a predefined class, such as "upregulated," "downregulated," or "unchanged," based on its expression profile. This differs from clustering, which simply groups similar genes without necessarily assigning them to predetermined categories. The Bayesian framework allows us to quantify the uncertainty associated with the classification of each gene, reflecting the confidence in the assignment. Suppose we're studying a disease and want to identify genes that are specifically associated with its onset. We might use a naive Bayes classifier, which models the probability of observing a specific gene expression profile given a particular class (disease or control). This provides a powerful tool for identifying genes implicated in the disease, even if their expression changes are subtle or complex. The inherent uncertainty quantification in the Bayesian approach helps us to discern which classifications are well supported and which remain uncertain, thus enabling us to focus our attention and experimental validation efforts.

The choice between Bayesian clustering and classification hinges on the research question. If the goal is to discover underlying patterns and structures within the data without pre-defined categories, then clustering is the preferred approach. However, if specific classes are already defined, and the task is to assign data points to those classes, then classification is the more appropriate method. A common scenario in microarray analysis involves both clustering and classification sequentially. We may first cluster genes based on their expression profiles, uncovering hidden relationships between genes, and subsequently classify those clusters into biologically meaningful categories, such as functional groups or pathways. Such a two-step approach helps refine and interpret the biological meaning derived from the analysis. For instance, we may cluster genes based solely on expression levels, then use that clustering result to inform a classifier that predicts the underlying biological function based on which cluster the genes belong to. This cascade method ensures that the interpretation of the data is both data-driven and biologically meaningful.

The practical application of Bayesian methods in microarray analysis involves careful consideration of prior distributions. The choice of prior can significantly influence the posterior distribution and, consequently, the clustering or classification results. Informative priors, incorporating existing biological knowledge, can improve the accuracy and interpretability of the results. However, improperly chosen informative priors can bias the results. Therefore, a thoughtful consideration of the available prior information and its potential impact is vital. It's crucial to strike a balance: overly informative priors might overshadow the data, leading to biased conclusions; on the other hand, non-informative priors might not leverage the power of existing knowledge, leading to less precise and potentially less informative results. In practice, exploring various prior choices and assessing their impact on the posterior is essential for robust and reliable analysis. The robustness of these Bayesian methods lies in their flexibility and the ability to explicitly model uncertainty. By quantifying uncertainty, we avoid overconfident interpretations, and allow for careful, reliable scientific conclusions. The Bayesian approach provides a framework that not only classifies or clusters the data but also reveals how confident we should be in these classifications or clusters. This level of detail is crucial for responsible interpretation of the biological results, as it enables a better understanding of the uncertainty inherent in the data and modeling process. Furthermore, the framework makes it straightforward to incorporate prior knowledge and evaluate the impact of such knowledge on our final analysis. Ultimately, a combination of thoughtful prior selection and careful consideration of the uncertainty estimates are vital for drawing reliable, scientifically-sound conclusions from microarray data.

5 Bayesian Approaches to Next-Generation Sequencing

Next-generation sequencing (NGS) generates massive datasets, presenting unique challenges for data analysis. Bayesian methods offer a powerful framework to address these challenges, leveraging prior knowledge and incorporating uncertainty inherent in the data. For instance, in variant calling, a central task in NGS analysis, Bayesian approaches excel by explicitly modeling the probabilities of different genotypes at each genomic position. This contrasts with frequentist methods which often rely on fixed thresholds, potentially leading to false positives or negatives. Bayesian methods, however, naturally incorporate uncertainty, providing posterior probabilities for each genotype, allowing researchers to more accurately identify true variants. Furthermore, Bayesian methods facilitate the integration of multiple data sources, like prior information on population allele frequencies and read quality scores. This holistic approach significantly enhances the accuracy and reliability of variant calls.

The power of Bayesian approaches extends to RNA-Seq analysis. This technique quantifies gene expression levels, providing critical insights into biological processes. Bayesian models provide a robust framework for handling the inherent complexities of RNA-Seq data, such as read mapping ambiguities and technical noise. By modeling the underlying biological process and incorporating technical variability, Bayesian approaches provide more accurate estimates of gene expression levels than many traditional methods. This refined accuracy enhances the reliability of downstream analyses, including differential expression studies and pathway analyses. The incorporation of prior knowledge, such as gene annotations or biological pathways, further enhances the precision and interpretation of RNA-Seq results. For example, we can use the prior knowledge of gene co-regulation to improve the accuracy of gene expression estimation.

Metagenomics, the study of microbial communities through NGS, also benefits tremendously from Bayesian methods. Analyzing complex microbial communities presents significant statistical challenges: large numbers of species, varying abundances, and high levels of sequencing error. Bayesian methods are adept at modeling this complex diversity. By explicitly modeling the probabilistic nature of species composition and accounting for sequencing errors, Bayesian approaches provide more accurate and robust estimates of species abundance and community structure. Moreover, the ability of Bayesian methods to integrate diverse sources of information, such as phylogenetic relationships between species or environmental covariates, can reveal intricate community dynamics often missed by other approaches. This allows researchers to get a much better understanding of the microbial landscape and their interrelation.

Bayesian methods offer several advantages over frequentist approaches in NGS analysis. They are particularly well-suited for handling uncertainty and incorporating prior knowledge, features that are crucial for navigating the complexity of NGS data. This leads to more accurate and reliable results, paving the way for deeper biological insights. However, implementing Bayesian methods requires careful consideration of model selection and computational resources. The choice of prior distributions and model structure significantly impacts the results, and a thorough understanding of Bayesian theory is vital for appropriate implementation. Choosing inappropriate priors can lead to inaccurate or biased inferences, while model complexity can impact the computational feasibility. Thus, researchers need to choose models that strike a balance between complexity and computational cost.

Despite the computational challenges, the benefits of Bayesian approaches in NGS analysis are significant. The enhanced accuracy, ability to integrate diverse data types and handle uncertainty make them an increasingly valuable tool for uncovering molecular processes and understanding biological systems at a deeper level. As NGS technology continues to evolve, generating even larger and more complex datasets, the role of Bayesian methods will only continue to grow, becoming an indispensable tool in the bioinformatician's arsenal. Furthermore, ongoing developments in computational methods are continually improving the efficiency and scalability of Bayesian techniques, making them accessible to a broader range of researchers and applications. The future of NGS data analysis is firmly rooted in the powerful probabilistic framework offered by Bayesian inference.

5.1 Variant Calling

Variant calling, a cornerstone of next-generation sequencing analysis, uses probabilistic models to identify genetic variations within a genome. Unlike simpler methods that rely on solely on read counts, Bayesian approaches leverage prior knowledge about mutation rates, base-calling error probabilities, and genomic context to enhance accuracy and reduce false positives. This refined approach is especially crucial when dealing with low-coverage sequencing data or complex genomic regions. For beginners, understanding the core concepts of Bayesian inference as applied to variant calling is paramount. We'll explore this process step by step, building upon the foundation laid in earlier chapters.

The process starts with aligning reads to a reference genome. This alignment step, typically performed using tools like BWA or Bowtie2, generates a set of aligned reads—each representing a short sequence of DNA. For each aligned position, we examine the observed bases from multiple reads to detect potential variants. The most common type of variant is a single nucleotide polymorphism (SNP), where a single nucleotide differs from the reference genome. Insertions and deletions (indels) are also commonly observed. These are variations involving the addition or removal of one or more nucleotides.

To call a variant, we use a statistical model to assess the probability that a true variation exists at that genomic position. This isn't a simple count of variant alleles; instead, we consider several factors. For example, the quality scores associated with each base call, reflecting the confidence in the sequencing technology's output, are crucial. Low-quality bases contribute less to the probability of a true variant. Furthermore, we must incorporate the sequencing error rate, because random errors can mimic true variations. The Bayesian approach formally incorporates these factors into a probabilistic framework.

Prior information significantly influences the posterior probability of a variant. For example, a prior distribution based on population-specific allele frequencies from databases like dbSNP can dramatically improve variant calling accuracy. Regions known to have high mutation rates might also have different priors than more conserved regions. The integration of such prior information into the model isn't optional; rather, it's integral to achieving precise variant detection. Imagine trying to identify a rare variation in a sparsely sequenced genome—prior information significantly increases the accuracy of the call.

A critical part of variant calling involves selecting an appropriate statistical model. Often, hidden Markov models (HMMs) or similar probabilistic graphical models are employed. These models capture the dependencies between neighboring bases and account for the correlation between sequencing reads. The choice of model depends on various factors, including the characteristics of the sequencing data, the type of variants being sought, and the available computational resources. The most sophisticated models incorporate complex dependencies and multiple sources of information. Simpler models might overlook crucial details, especially in regions with complex genomic architecture.

Many software packages are available for Bayesian variant calling, each with its own strengths and weaknesses. GATK's HaplotypeCaller, for example, uses a sophisticated HMM to model haplotypes, taking into account the possibility of multiple variants in a single region. Other tools like FreeBayes and VarScan also employ Bayesian approaches, although with variations in model specification and parameter estimation. Choosing the right software often involves balancing model complexity with computational efficiency and data-specific requirements. Remember to always thoroughly examine the documentation and relevant publications before applying the software to your own data.

Once the variant calling is complete, the results must be rigorously evaluated. Metrics such as sensitivity (the proportion of true variants correctly identified) and specificity (the proportion of true negatives correctly identified) are essential for assessing the performance. Receiver operating characteristic (ROC) curves and precision-recall curves can visualize the trade-off between sensitivity and specificity at different thresholds. Additionally, the variant calls can be compared against gold standard datasets, where available, to assess the accuracy of the results, especially for identifying false positives and false negatives.

Beyond SNPs and indels, Bayesian methods extend to detecting more complex variants, such as copy number variations (CNVs) and structural variations (SVs). These variations involve larger genomic rearrangements or changes in the copy number of specific genomic regions. Bayesian approaches offer flexible frameworks to model these more complex types of genetic alteration and thus advance our understanding of genome structural variability. Mastering the foundations of Bayesian variant calling empowers researchers to address these more advanced problems. Finally, remember that rigorous quality control and validation steps are essential before drawing conclusions from variant calls.

5.2 RNA-Seq Analysis

RNA sequencing (RNA-Seq) provides a powerful way to measure gene expression levels, offering a far more nuanced view than older microarray technologies. Unlike microarrays, which rely on pre-defined probes, RNA-Seq directly sequences the RNA molecules present in a sample, allowing for the detection of novel transcripts and isoforms. This depth of information is particularly valuable when investigating complex biological systems. Analyzing this data effectively, however, requires sophisticated statistical methods, and Bayesian approaches prove exceptionally well-suited to this task.

The inherent complexity of RNA-Seq data presents challenges. Read counts, representing the number of times a specific transcript is sequenced, are not directly proportional to the abundance of that transcript. Technical biases, such as differences in sequencing depth or library preparation, can also significantly influence the measured read counts. Moreover, we frequently encounter overdispersion, where the observed variance exceeds that predicted by a simple Poisson model. Bayesian methods address these challenges by explicitly incorporating uncertainty and prior information into the analysis.

A common task in RNA-Seq analysis is differential expression analysis—identifying genes whose expression levels differ significantly between two or more experimental conditions. Frequentist methods often rely on simplifying assumptions, such as equal variances, which may not hold true for RNA-Seq data. In contrast, Bayesian methods allow us to model the expression levels using more flexible distributions, such as the negative binomial, which accounts for overdispersion. We can also incorporate prior information about the expected expression levels or the variability between samples, leading to more robust and informative results. This flexibility becomes particularly crucial when dealing with datasets containing a small number of replicates.

Bayesian hierarchical models are particularly well-suited for RNA-Seq analysis. These models allow us to account for both technical and biological sources of variation. For instance, we can model the read counts for each gene as arising from a negative binomial distribution with parameters specific to that gene. These parameters, in turn, can be modeled as drawn from a higher-level distribution, reflecting the overall variability across genes. This hierarchical structure allows us to borrow information across genes, improving the precision of our estimates, especially for genes with low read counts. Incorporating prior information regarding the expected magnitude of differential expression further enhances the robustness of the analysis.

The choice of prior distributions is a crucial aspect of Bayesian RNA-Seq analysis. Informative priors, incorporating prior knowledge about gene expression patterns or experimental conditions, can significantly improve the accuracy and efficiency of the analysis, particularly when dealing with limited data. However, it's crucial to carefully consider the impact of prior selection, ensuring that the prior distribution reflects existing knowledge appropriately without overly influencing the posterior inferences. Prior elicitation techniques, using expert knowledge or data from previous experiments, play a pivotal role in obtaining informed priors.

Beyond differential expression, Bayesian methods find application in other aspects of RNA-Seq data analysis. Isoform quantification, determining the relative abundance of different transcript isoforms from the same gene, poses another challenging task. Bayesian methods can model the probabilistic assignment of reads to isoforms, providing more accurate estimates than simple count-based methods. Similarly, Bayesian approaches can be used for splice junction detection, identifying points in the RNA molecule where exons are spliced together. These analyses, often involving intricate models accounting for various sources of uncertainty, benefit greatly from the flexible framework offered by Bayesian inference.

Furthermore, exploring the entire landscape of gene expression using RNA-Seq data often necessitates the integration of information from multiple experiments or datasets. Bayesian methods, specifically Bayesian hierarchical models, excel at integrating data from disparate sources. By sharing information across experiments and accounting for their inherent variability, these models yield more accurate and reliable estimates. Ultimately, this integrated approach reveals a far richer understanding of the complex dynamics regulating gene expression. Therefore, mastering Bayesian methods is essential for anyone aiming to fully harness the vast potential of RNA-Seq data. The benefits extend to more accurate differential expression analysis, improved isoform quantification, and a comprehensive understanding of gene expression regulation. The flexibility and robustness of Bayesian techniques are invaluable assets in this increasingly data-rich field.

5.3 Metagenomics

Metagenomics offers a powerful lens through which we can examine complex microbial communities, revealing the intricate tapestry of life hidden within environmental samples. Unlike traditional microbiology, which focuses on cultivating individual organisms in the lab, metagenomics bypasses the cultivation step, enabling the direct study of microbial genomes from their natural habitats. This approach has revolutionized our understanding of microbial diversity, revealing a vast unseen world of organisms previously unknown to science. The sheer scale of microbial diversity, with estimates suggesting trillions of microbial species globally, only a tiny fraction of which are culturable, highlights the crucial role metagenomics plays in understanding the planet's biodiversity.

Analyzing metagenomic data presents unique challenges. The sheer volume of data generated from high-throughput sequencing technologies requires sophisticated computational approaches for efficient processing and analysis. The massive datasets generated often contain a mix of sequences from many different organisms, making assembly and annotation particularly demanding. Moreover, the presence of numerous unknown organisms complicates the task, often requiring the development of novel bioinformatic tools and algorithms. One example of this challenge is the necessity to deal with highly fragmented genomic sequences, since the initial sequencing output rarely represents complete genomes. The task is akin to reconstructing a jigsaw puzzle with many missing pieces, with the additional complexity that some pieces belong to puzzles that we have not even seen before.

Bayesian methods offer a particularly elegant approach to handling the complexities of metagenomic data analysis. They provide a robust framework for incorporating prior knowledge and handling uncertainty inherent in the data. For instance, a prior distribution reflecting the known phylogenetic relationships between microbial species can significantly improve the accuracy of taxonomic assignments in the analysis of the 16S rRNA gene marker. Furthermore, when analyzing the functional potential of a microbial community through analyzing protein-coding genes, Bayesian methods facilitate the probabilistic assessment of which metabolic pathways are present or absent. This probabilistic approach is crucial given the noisy and often incomplete nature of metagenomic data.

One of the key applications of Bayesian inference in metagenomics is taxonomic profiling. The process of identifying and quantifying the different microbial species present in a sample depends heavily on the comparison of short sequencing reads to established microbial databases. Bayesian approaches allow for a more accurate quantification of microbial abundance by considering the inherent uncertainty associated with sequence similarity, handling possible erroneous classifications with probabilistic calculations. Bayesian methods, like those employing Dirichlet priors in abundance estimation, are specifically effective in addressing this challenge, providing more accurate estimates of microbial community composition.

Furthermore, Bayesian methods are also useful for functional profiling of metagenomes. Here, the goal is to infer the functional capabilities of the microbial community based on the presence of specific genes. This involves predicting the metabolic pathways and other functions that the community can perform. Bayesian networks, with their ability to model complex relationships between genes and functions, are particularly well-suited to this task. The hierarchical nature of Bayesian inference facilitates integrating multiple layers of information, from individual genes to entire pathways and even ecosystem-level functions. This allows for a more holistic and comprehensive understanding of the community's functional potential.

Beyond taxonomic and functional profiling, Bayesian methods are applicable to other aspects of metagenomic data analysis. For example, in studies of the dynamics of microbial communities, these methods can be used to model the temporal changes in species abundance and functional composition, providing insights into community stability and succession. The ability of Bayesian inference to handle time series data and incorporate prior knowledge, for example, on the environmental factors influencing community composition, gives it a significant advantage over other statistical methods. Moreover, the probabilistic nature of Bayesian inference allows for uncertainty quantification regarding predictions regarding future community composition or function, offering more robust and reliable analyses.

In conclusion, Bayesian inference provides an effective and powerful toolkit for tackling the complexities of metagenomic data analysis. Its ability to handle uncertainty, incorporate prior knowledge, and model complex relationships makes it ideal for addressing many challenges in the field. From improving the accuracy of taxonomic profiling and functional predictions to modeling dynamic changes in microbial communities, Bayesian methods are poised to further advance our understanding of the unseen world of microorganisms and their pivotal roles in diverse ecosystems. The integration of Bayesian approaches into metagenomic pipelines is increasingly crucial for unlocking the full potential of this groundbreaking field. As sequencing technologies continue to improve and produce even larger datasets, the need for robust and sophisticated analytical approaches like Bayesian inference will only become more pronounced. It promises more detailed and accurate understanding of the rich and complex microbial world surrounding us.

6 Model Selection and Evaluation

Choosing the right model is crucial; a poorly chosen model can lead to inaccurate conclusions, regardless of how sophisticated the Bayesian methods are. Therefore, carefully evaluating your model's performance is just as important as building it. Several metrics help assess model fit. For example, the Deviance Information Criterion (DIC) balances model fit and complexity, penalizing overly intricate models that might overfit the data. Another commonly used metric is the Widely Applicable Information Criterion (WAIC), which offers a robust alternative to DIC, particularly useful when dealing with hierarchical models, common in genomic analyses. These metrics provide a numerical comparison, enabling you to rank different models objectively.

Remember that model selection isn't a one-size-fits-all process. The best approach depends on the specific biological problem you are addressing and the nature of your data. For instance, when analyzing gene expression data, you might find that a simpler model performs better than a highly complex one. This is because introducing unnecessary parameters can lead to overfitting and poor generalization to new datasets, rendering your conclusions unreliable in real-world applications. Conversely, for complex biological interactions, a more sophisticated model, such as a dynamic Bayesian network, may be necessary to capture the intricate relationships.

Cross-validation provides a powerful way to evaluate a model's predictive performance. This technique involves splitting your dataset into multiple subsets. The model is trained on a portion of the data and then tested on the remaining held-out portion. This process is repeated multiple times, using different subsets for training and testing in each iteration. The average performance across all iterations offers a robust estimate of how well your model generalizes to unseen data, a crucial aspect for ensuring the reliability of your conclusions. For instance, in phylogenetic inference, cross-validation helps determine the model's ability to accurately predict evolutionary relationships based on new sequence data.

K-fold cross-validation, a common approach, divides your dataset into 'k' equal-sized folds. Each fold serves sequentially as a test set, while the remaining k-1 folds constitute the training set. Leave-one-out cross-validation, a special case where k equals the number of data points, offers a more thorough but computationally expensive evaluation, especially valuable when dealing with limited datasets. The choice between these approaches depends on the size and characteristics of your dataset and your computational resources. Consider your time constraints when choosing the cross-validation method for your analysis.

Bayesian Model Averaging (BMA) provides a sophisticated way to address model uncertainty. Instead of selecting a single "best" model, BMA combines predictions from multiple models, weighted by their posterior probabilities. This approach acknowledges that no single model perfectly represents the underlying biological process; the truth likely lies somewhere in between several competing hypotheses. BMA's strength is its ability to account for this uncertainty, generating predictions that are more robust and less sensitive to the choice of a specific model. In microarray analysis, for example, BMA can improve the accuracy of differential gene expression estimates by incorporating information from various competing models representing different assumptions about gene expression patterns.

Implementing BMA can be computationally demanding, especially when dealing with many competing models. However, the benefits often outweigh the computational costs. The resulting predictions are more reliable due to the incorporation of model uncertainty. This makes BMA a valuable tool, particularly when analyzing complex biological systems where model uncertainty is significant. Furthermore, the posterior probabilities assigned to each model provide valuable insights into the relative support for different biological hypotheses, contributing to a more nuanced and scientifically rigorous interpretation of the results.

Remember, successful Bayesian inference in bioinformatics isn't solely about employing complex models; it's about choosing the right model, evaluating its performance rigorously, and interpreting the results within the context of the biological system under study. A simple model, well-evaluated and appropriately interpreted, can often yield more reliable conclusions than a complex model inadequately assessed. Prioritize understanding your data, selecting appropriate models, and rigorously evaluating their performance to generate meaningful and reliable insights into the intricate molecular processes that govern life. Always consider the limitations of your chosen model and the implications for the interpretation of your results, striving for transparency and thoroughness throughout your analysis. The pursuit of accurate modeling should always be guided by a deep understanding of the underlying biology and the principles of sound statistical inference. Approaching your analysis with critical thinking, meticulous attention to detail and a genuine curiosity for the data will elevate the quality and impact of your research.

6.1 Model Comparison Metrics

Comparing Bayesian models requires careful consideration of several metrics, going beyond simple accuracy. We're not simply picking the "best" model, but rather the model that best balances explanatory power with complexity, avoiding overfitting. A model that perfectly fits the training data might perform poorly on unseen data, reflecting a lack of generalizability. Therefore, we must evaluate how well a model predicts new, independent data, a crucial aspect often overlooked by newcomers.

One essential metric is the Deviance Information Criterion (DIC). DIC balances model fit with complexity. Lower DIC values generally indicate better models; it penalizes models with more parameters, thus discouraging overfitting. Think of it as a refined version of AIC (Akaike Information Criterion) specifically tailored for Bayesian models. Interpreting DIC involves comparing values across various models: a smaller DIC suggests superior performance. Remember, however, that DIC is not universally preferred, and its limitations have been discussed.

Another valuable tool is the posterior predictive p-value. This assesses how well a model predicts observed data given its own fitted parameters. This involves simulating new data sets from the posterior predictive distribution and comparing these simulations with your real data. A low posterior predictive p-value could suggest that the model isn't adequately capturing the underlying data generating process. A high p-value, however, doesn't automatically guarantee a good model, highlighting the importance of considering other metrics. It's a valuable check to see if there's something missing in your model.

The Log-Pseudo Marginal Likelihood (LPML) provides another way to compare models. The LPML assesses the overall predictive performance of a model. It averages the log-likelihood of the observed data across all possible parameter values, weighted by their posterior probabilities. Higher LPML values are associated with better predictive ability. This metric is especially helpful when dealing with complex models where simple metrics may not be sufficient. Keep in mind that the calculations involved can be computationally intensive, especially with high-dimensional models.

Beyond these, consider the Watanabe-Akaike Information Criterion (WAIC). WAIC offers an alternative approach to DIC, addressing some of DIC's limitations by directly estimating the out-of-sample predictive performance. WAIC effectively approximates the expected log pointwise predictive density, providing a comprehensive measure of predictive accuracy. It's less computationally intensive than LPML, particularly attractive in complex scenarios. Its strength lies in its robustness and ability to reliably estimate predictive performance.

Effective model comparison frequently involves a combination of these metrics. There's no single "best" metric; the optimal choice depends on the context, your specific goals, and the nature of the data. Relying solely on one metric could lead to misleading conclusions. A comprehensive analysis using multiple metrics provides a more nuanced and reliable assessment of competing models. Moreover, carefully considering your prior information about the biological system being modeled can guide your interpretation of these metrics.

Remember, visualization plays a key role. Visualizing posterior distributions, comparing predictions to observed data, and analyzing the model's behavior in different scenarios can offer invaluable insights, often complementing the quantitative results from the metrics. In addition to formal numerical metrics, always combine this with a qualitative analysis: does the model's structure align with what you know about the biological process? This holistic assessment greatly enhances the value and reliability of your inferences.

Furthermore, don't underestimate the importance of careful model specification. A poorly specified model, no matter how well it performs on the chosen metrics, will provide unreliable results. Ensure that the model accurately reflects your understanding of the biological system and that your prior distributions are sensible. A detailed examination of model assumptions is critical, particularly in the case of Bayesian methods, which directly incorporate prior knowledge. It's common for beginners to overlook this, leading to biased or less efficient results.

Finally, remember to consider the computational cost. Certain model comparison methods are computationally more demanding than others. The choice of metric should also reflect practical considerations, including your computational resources and time constraints. Prioritize clarity and interpretability in your choice. Avoid metrics that are extremely complex or require specialized software if simpler, readily understandable alternatives exist. A well-chosen metric must be a practical tool, enabling easy interpretation and clear communication of your results. In essence, the best approach is a thoughtful combination of quantitative assessments and qualitative considerations.

6.2 Cross-Validation Techniques

Cross-validation is a crucial technique for evaluating the performance of a Bayesian model, ensuring it generalizes well to unseen data. It's particularly important in bioinformatics where datasets can be limited and noisy, leading to overfitting if not carefully addressed. Simply put, we want a model that accurately reflects the underlying biological processes, not just the quirks of our specific dataset. Ignoring this can result in misleading conclusions and ultimately, ineffective applications. Remember, the goal isn't just to fit the data; it's to extract meaningful biological insights.

The most common approach is k-fold cross-validation. Imagine splitting your data into k equally sized subsets. One subset is held out as a test set, while the remaining k-1 subsets are used to train the model. We repeat this process k times, each time using a different subset as the test set. The final performance metric is the average performance across all k iterations. This provides a robust estimate of how well the model will perform on entirely new data. A higher k value, like 10-fold cross-validation, generally leads to more reliable estimates, but increases computational cost.

Choosing the right k is a balance between computational cost and the accuracy of the estimate. While a higher k sounds better in theory, the increase in computational cost might not always justify the improved estimation. Too low a k value, say 2-fold, can lead to a less stable and less representative performance estimate. Think of it like tasting a single spoonful of soup versus several. One spoonful might not give you a true flavor profile. Experimentation and prior experience often guide this decision, and literature specific to your application can further refine this.

Beyond k-fold, leave-one-out cross-validation (LOOCV) exists. This is an extreme case of k-fold, where k equals the number of data points. Each data point is used as a test set while the remaining data points train the model. While providing an extremely unbiased estimate, it's computationally expensive, especially for larger datasets. It's often preferred when you have a limited dataset and prioritizing a precise performance assessment is essential, even at the expense of computation time. Consider its practical implications carefully before implementation, especially if computational resources are a factor.

Another important aspect is the choice of performance metric. The best metric depends on the specific application. For example, in gene prediction, accuracy and sensitivity might be paramount, while in variant calling, the F1-score might be more appropriate given the potential for false positives and false negatives. You must carefully consider the relative importance of true positives, true negatives, false positives, and false negatives given the context of your study. Misinterpreting results based on inadequate metrics is all too common. Beyond the selection of the validation technique, remember the importance of proper data preprocessing. This can include handling missing values, normalizing data, and dealing with outliers. Failure to perform these preprocessing steps can significantly impact the accuracy of model evaluation and lead to biased results. Cleaning and preparing the data is a critical and often underestimated step that significantly improves the accuracy and reliability of any validation analysis. This also includes understanding potential biases in your dataset itself. This understanding can inform your choice of validation technique, and ensure your results are both valid and impactful. Consider the distribution of your data—is it normally distributed? Are there any hidden factors that might be influencing the performance?

Finally, remember that cross-validation is not a silver bullet. It provides a valuable estimate of model performance, but it's just one piece of the puzzle. Combine cross-validation results with other model selection criteria, such as the Bayesian Information Criterion (BIC) or Deviance Information Criterion (DIC), for a more comprehensive evaluation. Don't rely solely on one metric; use multiple lines of evidence to arrive at a conclusion. Understanding the strengths and limitations of different evaluation techniques is paramount to drawing valid and meaningful inferences from your analysis. Your findings should be robust and reflect a comprehensive approach to model selection. Ignoring this principle can lead to overconfident conclusions, and inaccurate results.

Remember, the goal is to build a model that accurately reflects the biological reality, not just fit to the training data. By understanding and carefully applying various cross-validation techniques, choosing appropriate performance metrics and acknowledging the limitations of statistical tests, you can increase your confidence in the inferences drawn from your Bayesian models and ultimately contribute to a deeper understanding of the biological processes under investigation. Thorough and careful validation techniques directly translate to more robust and meaningful conclusions in your bioinformatics research. This process ensures the reliability and impact of your studies. Every step is crucial for drawing accurate conclusions.

6.3 Bayesian Model Averaging

Bayesian Model Averaging offers a powerful solution to the ever-present problem of model uncertainty. Unlike selecting a single "best" model, which can lead to overconfidence and potentially biased results, BMA elegantly acknowledges the inherent uncertainty in model selection. It does this by weighting the predictions of multiple models according to their posterior probabilities, effectively averaging the predictions across the model space. This approach provides a more robust and accurate estimate, mitigating the risk associated with choosing a potentially incorrect model. The result is a prediction that incorporates the collective wisdom, so to speak, of various competing models.

This averaging isn't simply a democratic process of giving each model an equal say. Instead, BMA uses the posterior model probabilities, calculated from the data and prior knowledge, to assign weights to each model. Models supported strongly by the data receive higher weights, while poorly supported models contribute less to the overall prediction. This method implicitly accounts for model complexity and avoids overfitting by penalizing overly complex models that may fit the training data well but generalize poorly to new, unseen data. Essentially, BMA rewards models that strike a good balance between goodness-of-fit and parsimony, a key aspect of robust statistical modeling.

Imagine you're trying to model gene expression data. You might consider several different models, each incorporating different assumptions about the underlying biological processes and data generating mechanisms. Perhaps one model assumes a linear relationship between gene expression and a specific environmental factor, while another assumes a non-linear relationship. Yet another may include interactions between multiple factors. Choosing just one model risks overlooking important nuances of the system. BMA, however, provides a way to combine insights from all these models, yielding a more comprehensive understanding of gene regulation. The resultant prediction thus becomes less sensitive to the choice of a single, potentially inappropriate model.

Implementing BMA requires careful consideration of computational complexity. Evaluating the posterior probability of each model can be computationally intensive, especially when dealing with a large number of models. Approximation methods, such as Markov Chain Monte Carlo (MCMC) techniques, are often employed to overcome these challenges. While the computational cost might initially seem daunting, the enhanced accuracy and robustness that BMA offers often outweigh this cost, particularly when dealing with complex biological systems where model uncertainty is high. It's crucial to balance computational feasibility with the desire for improved accuracy; sophisticated strategies for model space exploration may be necessary in demanding scenarios.

Furthermore, the choice of prior distributions plays a critical role in BMA. The prior distributions reflect our prior beliefs about the relative plausibility of different models. Informative priors, based on prior biological knowledge or previous experiments, can enhance the accuracy and efficiency of BMA. However, the use of overly strong or inappropriate priors can bias the results, highlighting the importance of carefully selecting priors that reflect prior knowledge accurately and prevent unintended influence on the outcome. A sensitivity analysis investigating how the results vary with the choice of prior distributions is always a sound practice to verify the robustness of the approach.

The benefits of BMA extend beyond improved prediction accuracy. It also quantifies model uncertainty, providing valuable insights into the robustness of the conclusions. This is achieved through the calculation of posterior model probabilities and the associated model weights. This information is crucial for interpreting the results and understanding the extent to which inferences are sensitive to model assumptions. By explicitly acknowledging and quantifying model uncertainty, BMA enhances the transparency and reliability of bioinformatic analyses. This is particularly crucial in scenarios where incorrect model selection can have significant implications for downstream analyses and biological interpretations.

In the context of bioinformatics, BMA finds widespread application in areas like gene regulatory network inference, phylogenetic analysis, and comparative genomics. For example, in gene regulatory network inference, researchers may consider various network structures, each representing different interactions among genes. BMA can be used to combine predictions from these different network structures, obtaining a more reliable and robust estimate of the network. In the same way, for phylogenetic analysis, it can integrate multiple tree topologies, effectively hedging against uncertainty in the evolutionary relationships among species. It's not a simple matter of combining data from all different models: the essence of BMA lies in weighting these according to their posterior likelihoods, giving a precise measure of reliability to the resulting inferences.

In conclusion, Bayesian Model Averaging presents a powerful and versatile method for handling model uncertainty in bioinformatic applications. By acknowledging the inherent limitations of selecting a single model and weighting predictions based on posterior model probabilities, BMA enhances the robustness and reliability of analyses while quantifying the uncertainty associated with model selection. While computationally demanding in some cases, the benefits in terms of accuracy, robustness, and transparency often make it a highly valuable tool for researchers working with complex biological data, improving overall confidence in the outcome and giving a strong basis for further biological investigations. Mastering its application requires attention to computational efficiency and careful selection of prior distributions, but the rewards are well worth the effort.

7 Advanced Topics and Applications

Bayesian nonparametrics offer a powerful alternative to traditional Bayesian methods when dealing with the inherent complexity of biological data. Unlike parametric methods that assume a specific form for the underlying distributions, nonparametric approaches allow the data to inform the shape of the distribution. This flexibility is particularly beneficial in bioinformatics where the true distribution of biological phenomena is often unknown or highly complex. Dirichlet process mixture models, for example, can effectively model the diverse expression patterns of genes in microarray experiments without making restrictive assumptions about the number of clusters or the form of the gene expression distributions. Such flexibility allows for a more accurate representation of biological reality, potentially revealing subtle patterns missed by simpler models. Implementation often involves Markov Chain Monte Carlo (MCMC) methods, which can be computationally intensive, but advancements in algorithms and computing power are continually making these methods more accessible.

The computational demands of Bayesian inference, particularly when dealing with high-dimensional datasets common in bioinformatics, pose significant challenges. MCMC methods, while powerful, can be slow to converge and require careful tuning of parameters. Approximations like variational inference offer faster alternatives, albeit at the potential cost of accuracy. Developing efficient and scalable algorithms remains a critical area of research, particularly for the ever-increasing volumes of data generated by next-generation sequencing technologies. Optimization of existing algorithms and the exploration of novel computational approaches are essential to ensure the practical applicability of Bayesian methods to larger and more complex biological problems. Efficient use of parallel computing and advanced data structures are crucial for managing computational burdens.

Looking towards the future, several exciting directions promise to further advance the integration of Bayesian inference into bioinformatics. The increasing availability of large, diverse datasets, coupled with the development of more sophisticated computational tools, will enable the construction of increasingly detailed and accurate probabilistic models of biological systems. The combination of Bayesian methods with machine learning techniques, particularly deep learning, holds immense potential for analyzing complex biological relationships and predicting functional outcomes. For instance, Bayesian neural networks, capable of incorporating prior knowledge about biological systems into deep learning models, might prove highly beneficial. Furthermore, exploring the application of Bayesian inference to new challenges in bioinformatics, such as the analysis of single-cell data and the development of personalized medicine approaches, presents many new and exciting avenues for research. The continuing expansion of computational power and the development of more robust and efficient algorithms are essential for realizing the full potential of this interdisciplinary approach. The field is primed for major breakthroughs, pushing the boundaries of what we can understand and predict in the biological world. Careful consideration of model selection, validation, and interpretation is critical to translate these advances into meaningful biological insights.

7.1 Bayesian Nonparametrics

Bayesian nonparametrics offer a powerful alternative to traditional Bayesian methods when the underlying distribution's complexity is unknown or assumed to be highly flexible. Unlike parametric models that assume a fixed number of parameters, nonparametric models allow the complexity of the model to grow with the data. This adaptability is particularly valuable in bioinformatics, where the true underlying distributions of biological data are often unknown and may exhibit considerable heterogeneity. For instance, consider modeling gene expression levels across a population; a parametric approach might assume a normal distribution, which may be inadequate if the data exhibits multiple modes or heavy tails. A nonparametric approach, however, can automatically adapt to the true, more complex distribution.

One prominent example of a Bayesian nonparametric method is the Dirichlet process (DP). The DP is a stochastic process that generates probability distributions. It allows us to model an infinite mixture of simpler distributions, providing flexibility to capture intricate patterns. For a beginner, this might initially sound complex, but the intuition is straightforward: imagine modeling the distribution of protein lengths. Instead of committing to a specific distribution like the exponential, we can use a DP to model a mixture of many potential distributions, allowing for the potential presence of rare, exceptionally long, or short proteins without pre-defining those possibilities. This adaptability allows us to avoid making restrictive assumptions about data generation, leading to more robust and accurate models.

Practical application in bioinformatics is readily apparent. Imagine you're analyzing microbiome data. The number of different microbial species present is unknown, and it's unreasonable to assume a fixed number. A Dirichlet process mixture model (DPMM) can automatically discover the number of distinct species and estimate their abundances from the data, allowing for the discovery of novel species or unexpected community structures. Similarly, in analyzing gene expression data, DPMMs can identify subgroups of genes with similar expression profiles, even if the number of subgroups is unknown beforehand. This is a significant advance over traditional clustering techniques which require specifying the number of clusters beforehand.

Beyond DPMMs, other Bayesian nonparametric approaches include the Gaussian process, useful for modeling continuous functions, like those describing gene regulatory networks, or protein folding dynamics. Gaussian processes allow us to predict values at unseen data points, interpolating between observed values. In analyzing gene regulatory data, this means we can infer regulatory relationships for genes where direct measurements are lacking. However, it's critical to understand that these advanced methods require careful consideration of computational resources, as the complexity can increase considerably.

The computational challenges of Bayesian nonparametrics are significant. Inference often requires Markov chain Monte Carlo (MCMC) methods, which can be computationally expensive, especially with large datasets. Furthermore, model selection becomes more involved due to the infinite-dimensional nature of the parameters. Careful consideration of the choice of prior distributions is essential to prevent overfitting or underfitting. While sophisticated algorithms exist to handle these computations, a solid understanding of computational complexity is crucial for a successful implementation. Appropriate software packages and efficient coding practices are vital to mitigate these challenges.

Despite the computational difficulties, the flexibility and power of Bayesian nonparametrics in bioinformatics should not be underestimated. They offer a principled approach to deal with uncertainty in the data generating processes, avoiding the restrictive assumptions of parametric models. Their ability to adapt to the data's complexity makes them well-suited for biological problems, where the true distributions are rarely known or simple. As computational power continues to increase and new algorithms are developed, the application of Bayesian nonparametrics in bioinformatics will undoubtedly expand, enabling deeper insights into complex biological phenomena. Therefore, while they present computational challenges, the rewards in terms of improved model accuracy and a deeper understanding of complex biological systems are significant and warrant further exploration. This exploration might involve starting with simpler datasets, progressively increasing complexity as the user gains experience with the methods. The rewards for overcoming these hurdles are a more accurate understanding of biological mechanisms, leading to potentially better diagnoses and treatment of diseases.

As a beginner, understanding these methods thoroughly might require focusing initially on a specific area of application and then expanding your knowledge to incorporate a wider variety of methods as you become more adept at interpreting results and troubleshooting computational issues. Remember that the power of Bayesian nonparametrics lies not just in their mathematical elegance, but also in their ability to provide more accurate and nuanced insights into biological data, providing a valuable tool in the ongoing quest to unravel the complexities of life. The development of more user-friendly software and comprehensive tutorials will continue to reduce the barrier to entry, making these powerful techniques accessible to a broader range of researchers and scientists. The combination of theoretical understanding and hands-on practice will pave the way towards effective application.

7.2 Computational Challenges

Computational demands in Bayesian inference for bioinformatics can be substantial, particularly when dealing with large datasets common in genomics and proteomics. The complexity stems from the iterative nature of Bayesian methods, requiring repeated calculations to update posterior distributions. For instance, Markov Chain Monte Carlo (MCMC) techniques, while powerful, necessitate numerous iterations to achieve convergence, a process which can be computationally intensive, especially for intricate models with many parameters. This can translate to extensive runtime, potentially hindering the analysis of vast biological datasets. Efficient algorithms and optimized software are crucial for mitigating this.

Many Bayesian models used in bioinformatics, such as hidden Markov models and Bayesian networks, involve complex probability calculations. These calculations can become computationally intractable for large datasets or complex models due to the combinatorial explosion in the number of possible states or configurations. Approximation methods, like variational inference or expectation propagation, are often used to alleviate this issue, offering tradeoffs between computational speed and accuracy. The choice of approximation method often requires careful consideration based on the specific application and the desired level of accuracy.

Another significant hurdle is the availability of sufficient computational resources. High-throughput sequencing technologies generate massive datasets requiring considerable processing power and memory. Analyzing these data using Bayesian methods may necessitate high-performance computing clusters or cloud computing services, resources which can be expensive and potentially inaccessible to many researchers. Smart data management strategies and efficient algorithms become even more important in the face of these limitations. Efficient storage and retrieval of data, as well as the use of parallelization techniques, can help to mitigate this challenge.

Furthermore, model selection itself presents computational challenges. Comparing multiple Bayesian models requires evaluating their posterior probabilities, which can be computationally demanding when considering a large model space. Techniques like Bayesian model averaging, while statistically desirable, can significantly increase computational cost. Approaches to streamline model selection processes are therefore constantly being developed, incorporating strategies such as more efficient model exploration algorithms and the clever exploitation of model structure.

The choice of prior distribution also influences the computational complexity of Bayesian inference. Highly informative priors can simplify computations but may introduce biases. Conversely, less informative priors often necessitate more extensive computations to obtain accurate posterior inferences. Careful selection of priors, balancing informativeness and computational tractability, therefore requires substantial consideration. This demands a deep understanding of the biological system and the data at hand to inform this crucial choice in the modelling process.

Finally, effective visualization and interpretation of results from complex Bayesian models remains a challenge. High-dimensional posterior distributions can be difficult to visualize and interpret intuitively, requiring specialized visualization tools and techniques to effectively communicate findings. Developments in computational visualization are therefore crucial to unlocking the full potential of Bayesian methods in bioinformatics. This is a rapidly evolving area and ongoing research promises new tools to handle the complexity of visualization of results from Bayesian analyses. The challenge lies not only in computing the results but also in making them readily understandable for biologists.

7.3 Future Directions

The field of Bayesian inference in bioinformatics is poised for explosive growth. We've seen remarkable progress in applying Bayesian methods to increasingly complex biological datasets, but significant opportunities remain. One exciting area is the integration of diverse data types. Imagine combining genomic data, proteomic data, metabolomic data, and even clinical information within a single, unified Bayesian framework. This holistic approach could unlock previously hidden relationships and offer unprecedented insights into disease mechanisms, drug responses, and evolutionary processes. Such integrated analyses necessitate the development of sophisticated, scalable algorithms capable of handling massive datasets while maintaining computational tractability.

Furthermore, the development of more flexible and robust Bayesian models is crucial. Many current approaches rely on strong assumptions about the underlying data distributions, which may not always hold in reality. Exploring Bayesian nonparametrics, which allow the model's complexity to adapt to the data, is essential. This could lead to more accurate and reliable inferences, especially in scenarios involving complex interactions and high dimensionality. Simultaneously, advancements in computational statistics will be vital to handle the increased computational demands associated with these complex models. This may involve exploiting parallel processing capabilities or exploring new approximation techniques.

Another key direction lies in improving the interpretability of Bayesian models. While Bayesian inference provides probability distributions over model parameters, interpreting these distributions in a biologically meaningful way can be challenging. Developing improved visualization techniques and interpretable summaries of posterior distributions are needed to facilitate the communication of results to biologists with diverse backgrounds. This includes the creation of user-friendly software tools that simplify the model building process and present results in an accessible format. Clearer communication of uncertainty is also critical.

The increasing availability of large-scale biological databases presents both opportunities and challenges. Leveraging these databases effectively necessitates the development of new Bayesian methods capable of handling the inherent heterogeneity and noise present in these datasets. Data integration strategies become even more critical here, with the goal of developing a comprehensive understanding of complex biological systems. Careful consideration of bias, both in data collection and model design, will be essential to ensure robustness and generalizability of results.

Finally, the ethical considerations surrounding the use of Bayesian methods in bioinformatics cannot be ignored. As the field advances, it's crucial to establish ethical guidelines for the use and interpretation of Bayesian inferences, particularly in sensitive areas such as personalized medicine and genetic screening. Transparency in methods and careful consideration of potential biases are essential to prevent unintended consequences. Collaboration between bioinformaticians, biologists, and ethicists will be critical to navigate these complex issues effectively. The responsible application of Bayesian techniques requires thoughtful consideration of both scientific rigor and societal impact. The future holds tremendous potential; however, it is through addressing these challenges that we unlock its true promise.

www.ingramcontent.com/pod-product-compliance
Ingram Content Group UK Ltd.
Pitfield, Milton Keynes, MK11 3LW, UK
UKHW020157290926
15471UKWH00005B/914